FCC Rules and Regulations
for the Amateur Radio Service

Includes the complete Part 97 rules from Title 47 of the Code of Federal Regulations

---

Effective February 23, 2007

Now Including:
The FCC Rules and You

ARRL *The national association for AMATEUR RADIO*

225 Main Street, Newington, CT 06111-1494

Copyright © 2008-2011 by

The American Radio Relay League, Inc

Copyright secured under the Pan-American Convention

Printed in USA

Quedan reservados todos los derechos

ISBN: 978-0-87259-123-3
ARRL Order Number: 1173

Second Edition
Second Printing

Table of Contents

Legally, Safely Appropriately—The FCC and You

Subpart A — General Provisions
Section 97.1 Basis and purpose.
Section 97.3 Definitions.
Section 97.5 Station license grant required.
Section 97.7 Control operator required.
Section 97.9 Operator license grant.
Section 97.11 Stations aboard ships or aircraft.
Section 97.13 Restrictions on station location.
Section 97.15 Station antenna structures.
Section 97.17 Application for new license grant.
Section 97.19 Application for a vanity call sign
Section 97.21 Application for a modified or renewed license grant.
Section 97.23 Mailing address.
Section 97.25 License term.
Section 97.27 FCC modification of station license grant.
Section 97.29 Replacement license grant document.

Subpart B — Station Operation Standards
Section 97.101 General standards.
Section 97.103 Station licensee responsibilities.
Section 97.105 Control operator duties.
Section 97.107 Reciprocal operating authority.
Section 97.109 Station control.
Section 97.111 Authorized transmissions.
Section 97.113 Prohibited transmissions.
Section 97.115 Third-party communications.
Section 97.117 International communications.
Section 97.119 Station identification.
Section 97.121 Restricted operation.

Subpart C — Special Operations
Section 97.201 Auxiliary station.
Section 97.203 Beacon station.
Section 97.205 Repeater station.
Section 97.207 Space station.
Section 97.209 Earth station.
Section 97.211 Space Telecommand station.
Section 97.213 Telecommand of an amateur station.
Section 97.215 Telecommand of model craft.
Section 97.217 Telemetry.
Section 97.219 Message Forwarding System.
Section 97.221 Automatically controlled digital station.

Subpart D — Technical Standards
Section 97.301 Authorized frequency bands.
Section 97.303 Frequency sharing requirements.
Section 97.305 Authorized emission types.
Section 97.307 Emission standards.
Section 97.309 RTTY and data emission codes.
Section 97.311 SS emission types.
Section 97.313 Transmitter power standards.
Section 97.315 Certification of external RF power amplifiers.
Section 97.317 Standards for certification of external RF power amplifiers.

Subpart E — Providing Emergency Communications
Section 97.401 Operation during a disaster.
Section 97.403 Safety of life and protection of property.
Section 97.405 Station in distress.
Section 97.407 Radio amateur civil emergency service.

Subpart F — Qualifying Examinations Systems
Section 97.501 Qualifying for an amateur operator license.
Section 97.503 Element standards.
Section 97.505 Element credit.
Section 97.507 Preparing an examination.
Section 97.509 Administering VE requirements.
Section 97.511 Examinee conduct.
Section 97.513 VE session manager requirements
Section 97.515 [Reserved]
Section 97.517 [Reserved]
Section 97.519 Coordinating examination sessions.
Section 97.521 VEC qualifications.
Section 97.523 Question pools.
Section 97.525 Accrediting VEs.
Section 97.527 Reimbursement for expenses.

Appendices
Appendix 1 Places Where the Amateur Service is Regulated by the FCC
Appendix 2 VEC Regions

Legally, Safely, Appropriately —The FCC Rules and You

Dan Henderson, N1ND

Legally — Safely — Appropriately. These are the three hallmarks that allow each radio amateur to fully enjoy their on-the-air experience.

It is the responsibility of each amateur to operate their station within the rules governing the Amateur Radio Service (legally). The second responsibility is to take precautions to ensure that their activity doesn't pose harm to themselves or others (safely). And finally, each amateur is charged with making sure that their activities on the air are in keeping with the operating standards set by the rules, as well as those standards that have developed within the amateur community over time (appropriately).

THE AMATEUR RADIO SPECTRUM

The electromagnetic spectrum is a limited resource. Every kilohertz of the radio spectrum represents precious turf that is blood sport to those who lay claim to it. Fortunately, the spectrum is a resource that cannot be depleted — if misused, it can be restored to normal as soon as the misuse stops. Every minute of every hour of every day, we have a fresh chance to use the spectrum intelligently.

Although the radio spectrum has been used in a certain way in the past, changes are possible. Needs of the various radio services evolve with technological innovation and growth. There can be changes in the frequency bands allocated to the Amateur Service, and there can be changes in how we use those bands.

Amateur Radio is richly endowed with a wide range of bands starting at 1.8 MHz and extending above 300 GHz. Thus, we enjoy a veritable smorgasbord of bands with propagation "delicacies" of every type. To our benefit, radio propagation determines how far a signal can travel. Specific frequencies may be reused numerous times around the globe.

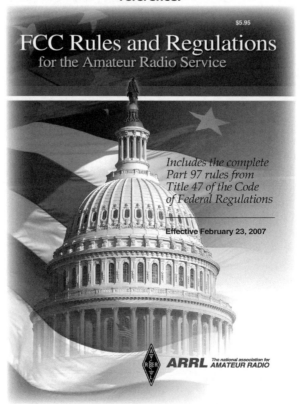

A current copy of the FCC rules is a must for every station. This booklet is a great way to keep the rules at your fingertips for easy reference.

WHERE DO THE RULES COME FROM?

International Regulation of the Spectrum

Amateur Radio frequency band allocations don't just happen. Band allocation proposals must first crawl through a maze of national agencies and the International Telecommunication Union (ITU) with more adroitness than a computer-controlled mouse. Simultaneously, the proposals affecting Amateur Radio have to run the gauntlet of competing interests of other spectrum users.

Treaties and Agreements

To bring some order to international relationships of all sorts, nations sign treaties and agreements. Otherwise (with respect to international communications) chaos, anarchy and bedlam would vie for supremacy over the radio spectrum. Pessimists think we already have some of that, but they haven't any idea how bad it could be without international treaties and agreements.

International Telecommunication Union

The origins of the ITU trace back to the invention of the telegraph in the 19th century. To establish an international telegraph network, it was necessary to reach agreement on uniform message handling and technical compatibility. Bordering European countries worked out some bilateral agreements. This eventually led to creation of the ITU at Paris in 1865 by the first International Telegraph Convention, which yielded agreement on basic telegraph regulations.

Plenipotentiary Conferences

The ITU has Plenipotentiary Conferences every four years. A "Plenipotentiary" is a conference that is fully empowered to do business. The conferences determine general policies, review the work of the Union and revise the Convention if necessary. The conferences also elect the Members of the Union to serve on the ITU Council, and elect the Secretary-General, Deputy Secretary-General, the Directors of the Bureaus and members of the Radio Regulations Board.

World Radiocommunication Conferences

ITU World Radiocommunication Conferences (WRCs) are held every two years with agendas agreed at the previous WRC and confirmed by the ITU Council. Various issues related to Amateur Radio may come up at these conferences. For example, amateur frequency allocations may be made or modified (such as the creation of the 30, 17 and 12 meter bands in 1979). In addition, changes made to the primary allocations of another service may affect the secondary status of Amateur Radio allocations.

Inter-American Telecommunication Commission (CITEL)

CITEL is the regional telecommunication organization for the Americas. It is a permanent commission under the Organization of American States (OAS), with a secretariat in Washington, DC. The CITEL Assembly meets every four years, while its committees meet one or more times yearly. ARRL and IARU Region 2 are active participants in Permanent Consultative Committee no. 3 (PCC.III — Radiocommunications).

TELECOMMUNICATIONS REGULATION WITHIN THE UNITED STATES

Practically from the day you started studying for your first Amateur Radio license, you encountered two terms that brought home the significance of the responsibilities that accompany an Amateur Radio license. Those two terms are FCC and Part 97.

FCC of course refers to the Federal Communications Commission, the agency of the US government responsible for making and enforcing the rules for the Amateur Radio Service. This five-person commission is charged with writing the rules that govern the licensing process for the various radio services in the US — commercial and private. The FCC oversees services as diverse as broadcast television and radio; communications services, such as wireless technologies; and wired services, such as cable television and telephone carriers. The FCC is charged with determining the standards for obtaining a federal license for the various services, as well as determining the privileges and technical standards that must be adhered to by each licensee.

Assisting the Commissioners in this task are hundred of employees. These experts help develop the standards and criteria for licensing and for operating under a Commission license. They also handle enforcement issues when rules are violated.

World Radiocommunication Conferences (WRCs), conducted by the ITU, are held every few years to amend the international Radio Regulations. At the 2003 conference, after years of preparatory work we were successful in achieving a realignment of 40 meters to increase the worldwide Amateur Radio allocation to 7000-7200 kHz effective in 2009. Initial debates on the realignment were held in this Sub Working Group at the conference and the proposal worked its way through the process to become reality.

Code of Federal Regulations

The full FCC rules are found in *Code of Federal Regulations* (*CFR*). The 50 "titles" of the CFR contain the laws and regulations of the United States. Those laws and regulations dealing with telecommunications are found in Title 47 of the CFR.

Title 47 is further divided into three general subsections, each including additional sections known as "Parts." The three main subsections include the rules and responsibilities for the three units of the federal government that are responsible for some area of national telecommunication policy. The first subsection, which includes Parts 0 through 199 are the rules for the Federal Communications Commission. The remaining two subsections are the regulations for the Office of Science and Technology Policy and National Security Council (under direction of the White House), and for the National Telecommunications and Information Administration (NTIA), under the Department of Commerce.

The functions relating to assignment of frequencies to radio stations belonging to, and operated by, the United States Government are assigned to the Assistant Secretary of Commerce for Communications and Information (Administrator, NTIA). Among other things, NTIA:

1) Coordinates telecommunications activities of the Executive Branch;

2) Develops plans, policies and programs relating to international telecommunications issues;

3) Coordinates preparations for US participation in international telecommunications conferences and negotiations;

4) Develops, in coordination with the FCC, a long-range plan for improved management of all electromagnetic spectrum resources, including jointly determining the National Table of Frequency Allocations; and

5) Conducts telecommunications research and development.

Obviously, the FCC and the NTIA work closely together on many issues relating to radio spectrum. Many Amateur Radio bands, especially the UHF and higher frequency bands, are shared with a variety of government agencies. As long as we remain good sharing partners, these agencies can be powerful advocates for protecting Amateur Radio frequencies from other users.

For the most part, the rules for the Amateur Radio Service are contained in CFR Title 47, Part 97. That's why the term "Part 97" is used when referring to the rules for Amateur Radio. It is important to note that sections of Title 47 other than Part 97 do impact the Amateur Radio Service in some way. For example, Part 2 contains specific information on geographic areas where amateur operations are restricted, such as power limitations around certain US military sites. Part 15 contains standards for unlicensed low-power devices that could cause potential interference to amateur bands. The best resource to view the current versions of these related parts is online at **ecfr.gpoaccess.gov**. This is the official online version of the CFR, maintained by the National Archives.

Why We are Here

The five basic principles for the Amateur Radio Service are clearly established at the very beginning of the rules:

§97.1 Basis and purpose.

The rules and regulations in this part are designed to provide an amateur radio service having a fundamental purpose as expressed in the following principles:

(a) Recognition and enhancement of the value of the amateur service to the public as a voluntary noncommercial communication service, particularly with respect to providing emergency communications.

(b) Continuation and extension of the amateur's proven ability to contribute to the advancement of the radio art.

(c) Encouragement and improvement of the amateur service through rules which provide for advancing skills in both the communication and technical phases of the art.

(d) Expansion of the existing reservoir within the Amateur Radio service of trained operators, technicians, and electronics experts.

(e) Continuation and extension of the amateur's unique ability to enhance international goodwill.

Though often referred to as a hobby, Amateur Radio crosses many boundaries. The licensees of the Amateur Service play important roles in public service and emergency communications as well as the development of new and innovative technological advancements. They function as trained communicators available to assist others, but they also participate to enjoy camaraderie and fun with the new friends they meet on the airwaves. Some amateurs seek new operating challenges or pursue hands-on opportunities to learn new things about radio or electronics. The rules recognize that amateurs have distinct talents that they make available to serve their communities.

OVERVIEW OF THE PART 97 SUBPARTS

Part 97 is divided into six distinct subparts, A through F. Each subpart deals with rules and content in specific areas. Taken together, the subparts comprise the knowledge each licensee needs to legally, safely and appropriately operate on the air. Let's start with a brief overview of the parts and then jump into the details of each one.

Subpart A deals with the broad *general provisions* of what an individual is required to do when operating. It also includes specific definitions (§97.3) of various terminology used through the rules. Subpart A covers license grants, call signs and general requirements for becoming a Commission licensee. It also includes certain requirements and protections regarding your station location and antennas.

Subpart B discusses *station operation standards*. These include the responsibilities of the licensee and the control operator of any amateur station. Subpart B details permitted and prohibited transmissions, proper station identification, third party and international communications, and restricted operations.

Amateur stations are allowed special types of operations, and these are covered in **Subpart C**. Included in this portion of the rules are the regulations that govern how some of the most popular methods of amateur communications are carried out. At some point, almost every amateur will be involved with one or more of these *special operations*. They include basic repeater operation or message forwarding systems, as well as more "exotic" operations such as satellite communications or remote control of amateur stations.

The *technical standards* of how we operate are found in **Subpart D**. Each amateur is responsible for the technical quality of their transmissions and for making sure that signals are transmitted only on frequencies authorized by their license. The frequency allotments, signal emission standards and power limitations for each US license class can be found here.

As one of the basic purposes of the Amateur Radio Service, each licensee has a role to play in *supporting and providing emergency communications*. The guidelines for this, found in **Subpart E,** are the reason many of today's licensees got involved.

Unlike some other radio services authorized by the FCC, each person seeking to become an amateur operator must pass one or more written examinations. **Subpart F** delineates the *qualifying examination systems* through which individuals may obtain and upgrade their operating privileges.

Don't be confused or overwhelmed by the scope or magnitude of Part 97. Though some areas may be a bit complex or difficult to understand, the six subparts work together to provide the basic framework for each amateur to operate legally, safely and appropriately.

SUBPART A — GENERAL PROVISIONS
Your Amateur Radio License

Your journey into Amateur Radio begins with a single important piece of paper — your license. This grant from the FCC is actually two licenses in one — your station license and your operator license. This document is your authorization to transmit on the amateur bands and it conveys your operating privileges.

An individual who becomes a Commission licensee is granted one — and only one — operator/station license. To qualify, you must pass one or more written examinations administered by a team of Volunteer Examiners (VEs) who conduct exams under the auspices of an accredited Volunteer Examiner Coordinator (VEC). **Table 1** gives an overview, and exam requirements are covered in more detail later in this chapter.

Table 1
Currently Issued Amateur Operator Licenses[†]

Class	Written Examination	Privileges
Technician	Technician-level theory and regulations. (Element 2)*	All amateur privileges above 50.0 MHz. Some CW privileges on 80, 40, 15, 10 meters. Some SSB and data privileges on 10 meters.
General	Technician and General theory and regulations. (Elements 2 and 3)	All amateur privileges except those reserved for Advanced and Amateur Extra.
Amateur Extra	Technician and General theory, plus Extra-class theory (Elements 2, 3 and 4)	All amateur privileges.

[†]A licensed radio amateur must pass only those elements that are not included in the examination for the amateur license currently held. Novice and Advanced licenses are no longer being issued but may be renewed. No Morse code test is required for any license.
*If you hold an expired Technician license issued before March 21, 1987, you can obtain credit for Element 3. You must be able to prove to a VE team that your Technician license was issued before March 21, 1987 to claim this credit.

Anyone, except for a representative of a foreign government, is eligible to hold a US amateur license. You must have a valid US mailing address, though [§97.5(b)(1)]. Remember that it is your responsibility to keep your mailing address current with the FCC. Failure to do so can result in the suspension or revocation of your license [§97.23].

Your station license designates your call sign — a prized possession for most hams. When you receive your initial license, you will be assigned a call sign from the Sequential Call Sign Assignment System. Many amateurs choose to participate in the Vanity Call Sign program [§97.19], which allows licensees to pick their own call sign following the guidelines for their license class. More on call signs in a bit.

Keep in mind that a station license — your call sign — is only an identifier. A station license does not convey any operator privileges. All operating privileges come to you through the license class you have earned through your examinations.

The FCC maintains all licensee information in a central database known as ULS, the Universal Licensing System. As soon as your information appears in the ULS you are considered licensed and may operate using your assigned call and privileges. When you are issued your license, or when you renew (if you have renewed since December 3, 2001), you are also issued a Federal Registration Number (FRN). Any modification, renewal or upgrade of a license must includes the licensee's FRN, which in most cases for newer licenses is printed on the license itself. You can also find your FRN using the FCC ULS online search at **www.fcc.gov/wtb/uls** or ARRL's call sign lookup, available at **www.arrl.org**.

License Renewals and Updates

The FCC prefers that you renew your license or make modifications (update information such as your address) online using ULS. It's still possible to conduct business with the Commission "the old fashioned way" — by filing paper forms — but you should know the following:

1) ARRL members may use NCVEC Form 605 for everything except requests for new vanity call signs. This is not an FCC form, but it is accepted by the FCC if it is processed by an accredited VEC. You can download a copy from **www.arrl.org/fcc/forms.html**. The ARRL VEC processes renewals or modifications via NCVEC Form 605 as a free membership service. Amateurs may also deal directly with the FCC using FCC Form 605, found at **www.fcc.gov/Forms/Form605/605main.pdf**.

2) ARRL members using the NCVEC Form 605 should send it to the ARRL VEC, 225 Main St, Newington, CT 06111. Do not send NCVEC Form 605 to the FCC; it won't be accepted. Amateurs using the FCC Form 605 must send it to the FCC, 1270 Fairfield Rd, Gettysburg, PA 17325. There is no fee for renewing your amateur license at this time, unless your call sign was issued under the Vanity Call Sign Program.

3) The FCC will not process renewal applications received more than 90 days prior to the license expiration date.

4) It's a good idea to make a copy of your renewal application as proof of filing before your expiration date. If your application is processed before the expiration date, you may continue to operate until your new license arrives. Otherwise, you may not operate until your renewal is processed.

5) If your license has already expired, it is still possible to renew since there is a two-year grace period. If it has been expired more than two calendar years, you will have to take the current license test(s) to regain your amateur privileges.

6) You are required to keep your mailing address up to date. The FCC may cancel or suspend a license if their mail is returned as undeliverable.

For more information, see **www.arrl.org/arrlvec/**. You can also contact the VEC staff or Regulatory Information Branch at ARRL HQ for information on ULS and keeping your license current.

Call Sign Structure

The International Telecommunication Union (ITU) is an international organization that, among other things, sets the standards for the prefixes and formation of call signs in the various radio services worldwide. According to the ITU Radio Regulations, an amateur call sign must start with one or two letters as a prefix, although sometimes the first or second character might be a number (as in 8P for Barbados). The prefix is followed by a number indicating a call sign district (or "call area"), and then a suffix of not more than three letters. You can find the current allocated prefixes for each country online at **www.arrl.org/awards/dxcc/itucalls.html**.

Table 2
FCC-Allocated Prefixes for Areas Outside the Continental US

Prefix	Location
AH1, KH1, NH1, WH1	Baker, Howland Is.
AH2, KH2, NH2, WH2	Guam
AH3, KH3, NH3, WH3	Johnston Is.
AH4, KH4, NH4, WH4	Midway Is.
AH5K, KH5K, NH5K, WH5K	Kingman Reef
AH5, KH5, NH5, WH5 (except K suffix)	Palmyra, Jarvis Is.
AH6,7 KH6,7 NH6,7 WH6,7	Hawaii
AH7, KH7, NH7, WH7	Kure Is.
AH8, KH8, NH8, WH8	American Samoa
AH9, KH9, NH9, WH9	Wake Is.
AH0, KH0, NH0, WH0	Northern Mariana Is.
AL0-9, KL0-9, NL0-9, WL0-9	Alaska
KP1, NP1, WP1	Navassa Is.
KP2, NP2, WP2	Virgin Is.
KP3,4 NP3,4 WP3,4	Puerto Rico
KP5, NP5, WP5	Desecheo

Every US Amateur Radio station call sign is a combination of a 1 or 2-letter prefix, a number and a 1, 2 or 3-letter suffix. The first letter of every US Amateur Radio call sign will always be K, N or W, or from the AA-AL double letter block. Some examples: KA9XYZ, NN1N, WB2OSQ, KN4AQ, AB6ZZ, W3ABC. Certain prefixes are reserved for stations that are under FCC jurisdiction but located outside the continental US. Common examples are KP2 for the US Virgin Islands or KH6 for Hawaii. See **Table 2** for details.

The mailing address that you use when obtaining your initial sequential call sign determines the number used. The 10 call sign districts for the continental US are illustrated in **Fig 1**. So the call sign W1AW indicates a US station (from the prefix W). The call sign was initially assigned to that station while it was located in the first US call area (indicated by the number 1). The letters AW comprise the assigned suffix, completing the identification of that station.

Vanity Call Signs

Under the Vanity Call Sign Program, individual amateurs and club stations may select a distinctive call sign to replace the one assigned by the sequential system. There are many reasons for doing so. Some operators want shorter call signs, or call signs that are easy to say or send. A popular choice is a suffix with the initials of your name. Others may want to recapture a former call sign or that of a departed friend or family member. Some operators just grow tired of their current call sign and want to try something new.

Although there was a period in the 1970s when Amateur Extra licensees could select "1 × 2" call signs (N2CQ, K8MR and so forth), the current Vanity Call Sign Program is much broader. Beginning in 1996 amateurs could select any available call sign valid for their license class for a fee. The exact amount is subject to change each year, but it has always been reasonable considering that the term is 10 years. In addition to paying the fee when the Vanity Call Sign is issued, note that you must pay whatever fee is in effect at your scheduled license renewal. The number in a vanity call sign does not have to indicate the call area in which you reside.

Changing Locations

Over time, you may move to another state and find that the number in your call sign no longer matches the district of your current residence. Not a problem. You have the option of keeping your call sign even though you have changed districts, and many amateurs choose to do just that. At one time you were required to turn in your old call sign for a new one that matched your current district, but the FCC dropped this rule in the 1970s.

It's routine to hear stations on the air using call signs that do not correspond to their locations. You are not required by Part 97 to identify your station as portable or mobile, fixed or temporary, but it is certainly allowed and sometimes helps listeners understand your location. For example, N1ND operating in North Carolina might send N1ND/4 on 6 meters to help others point their beams in the right direction.

You will occasionally hear a station identify as "marine mobile" or "maritime mobile" This simply indicates that the station is being operated from onboard a boat or ship. Finally you may hear a station identify as "aeronautical mobile" indicating operation from on board an airplane. While these are legal IDs, remember that operation onboard a ship or airplane must be approved by the captain of the vessel or pilot of the plane.

Antennas and Support Structures

A growing concern for many hams is the placement and location of Amateur Radio antennas and supporting structures, typically towers. Amateurs understand that the effectiveness of their stations and communications ability centers on the type and quality of their antennas. Not everyone views antennas the same way, and amateurs occasionally are drawn into disputes with the neighbors and town officials over the height and placement of antennas and support structures.

In the 1980s, the FCC recognized that amateurs sometimes need assistance in dealing with local governments and zoning boards to pursue the privileges granted by their FCC license. From this was born a powerful tool — known as PRB-1 — that grants amateurs a limited federal preemption of local zoning ordinances and regulations. The purpose is to protect amateurs from outright prohibitions or unreasonable restrictions placed on antenna support structures.

PRB-1 has been incorporated into §97.15(b) and states: "Except as otherwise provided herein, a station antenna structure may be erected at heights and dimensions sufficient to accommodate amateur service communications. [State and local regulation of a station antenna structure must not preclude amateur service communications. Rather, it must reasonably accommodate such communications and must constitute the minimum practicable regulation to accomplish the state or local authority's legitimate purpose. See PRB-1, 101 FCC 2d 952 (1985) for details.]"

In addition to being part of the FCC rules, as of October 2007 a total of 25 states have incorporated PRB-1 protections into state as well. You can find complete information on PRB-1 and links to the specific pieces of state legislation adopting PRB-1 at **www.arrl.org/FandES/field/regulations/PRB-1_Pkg/index.html**

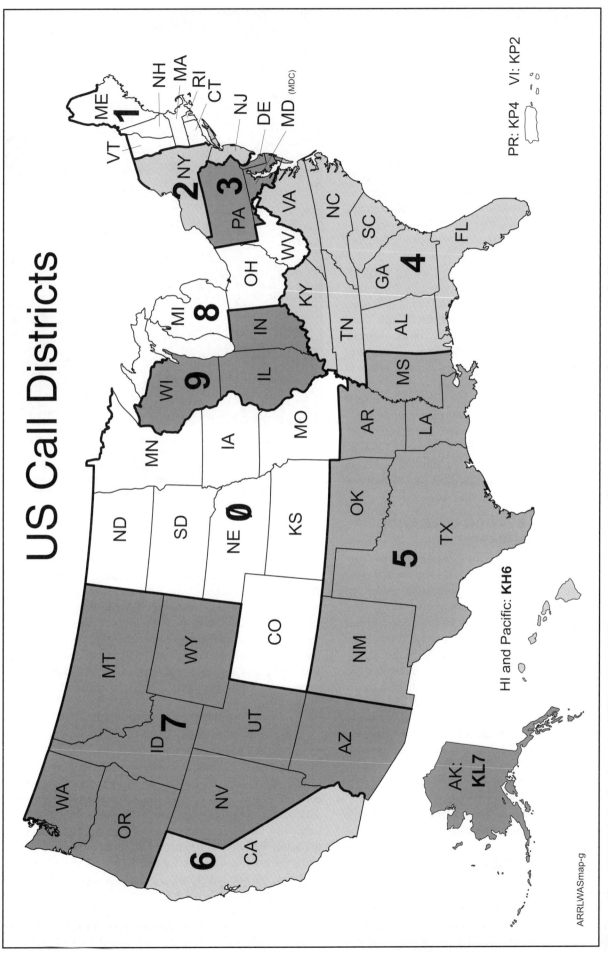

Fig 1 — The 10 US Call Districts established by the FCC. In most cases, the number in a US amateur call sign indicates where the operator lived when the FCC first issued that call sign. Alaska is part of the seventh call district, but has its own set of prefixes: AL7, KL7, NL7 and WL7. Hawaii is part of the sixth call district but also has its own set of prefixes: AH6, KH6, NH6 and WH6.

There are two things to remember as far as PRB-1 goes. First, PRB-1 does not give amateurs *carte blanche* to do what they wish in regard to antennas and support structures. PRB-1 requires local authorities to work with the amateurs to satisfy communications needs with the minimum amount of regulation needed to meet the legitimate needs of the local authority. Local regulators may not reject *all* amateur antennas, but they don't have to allow a structure either of a certain size, either.

The second important item to remember is that at this time PRB-1 applies only to regulation by local government officials. Many amateurs face similar (and sometimes more stringent) restrictions when dealing with what are known as CC&Rs — covenants, conditions and restrictions. Sometimes called simply "deed restrictions," CC&Rs are restrictions that are placed on property and deed, usually by the builder of a housing development or by a homeowner's association.

CC&Rs typically include limitations or guidelines for property appearance and use and may require approval of things such as site plan, building design and colors, placement of fences, landscaping and so on. Of concern to amateurs, CC&Rs often limit or prohibit outdoor antennas or supports. You should make it a point to find out about CC&Rs before entering into a real estate contract. They run with the property and you will have to abide by them once you're the owner.

The FCC has ruled that PRB-1 does not apply to CC&R situations. CC&Rs are private contacts, not public law or policy implemented by state or local government. For a good overview of the differences between PRB-1 issues and CC&R issues, see "What Should I Do Now?" from May 2007 *QST* and available online at **www.arrl.org/FandES/field/regulations/hender.pdf**.

SUBPART B — STATION OPERATION STANDARDS

When teaching license classes or speaking about my passion, I often say that your license is only the *second* most important piece of paper you receive in Amateur Radio. Of course you must have a license to get on the air, but I believe that your first QSL card (a written confirmation from another station that you have completed a two-way contact with them) is *the* most important "document" you receive.

Why do I believe that? Your first QSL card shows that you not only learned what you needed to pass the license exam, but also put that knowledge to use, got on the air and actively participated. Getting on the air is the point of the exercise, right? To do this successfully, Subpart B — Station Operation Standards — is critical to your pursuing Amateur Radio legally, safely and appropriately.

Thou Shall Not...

Front and center in Subpart B, §97.101(a) may well be the most important rule of all because it lays out the overriding responsibility of each amateur licensee: "In all respects not specifically covered by FCC Rules each amateur station must be operated in accordance with good engineering and good amateur practice."

This simple sentence underscores the responsibility of each operator to use good judgment and common sense when pursuing their interests. It means each licensee must continue

Getting permission from local government to put up antennas and antenna supports can be a problem in some areas. FCC's PRB-1 grants limited preemption of local zoning regulations and requires reasonable accommodation of amateur antennas.

to learn how stations interact properly, cooperate with fellow amateurs and treat each operator on the air with respect.

In many ways Subpart B is a laundry list of "do's and don'ts" that enable hundreds of thousands of amateurs on the air to share our limited bandwidth. Take the time to familiarize yourself with Subpart B. By following the letter and spirit of these rules, you can maximize your enjoyment while reducing potential problems or complaints on the air. A considerate operator will apply the rules and standards found in this section to his or her daily operation.

Sharing Our Spectrum

One of the most common hassles we experience on the air is crowded conditions. For example, trying to find a clear frequency on 20 meter SSB on a busy weekend morning can be a challenge. Effective spectrum use requires the cooperation of each and every operator. That's equally true during years of low solar activity when poor high-band propagation drives everyone to the low bands, or during years of high solar activity when the bands are filled with hams coming out of the woodwork to enjoy the great conditions. Remember we all share these frequencies: No one individual or group is assigned a frequency for exclusive use [§97.101(b)].

Over the years, to promote the orderly sharing of frequencies, some "gentleman's agreements" have emerged. Known as *voluntary band plans*, these are not hard and fast FCC rules, but rather guidelines for what type of amateur activity should take place in various parts of the spectrum. For example, a station trying to work the popular PSK31 digital mode would want to operate around frequencies where that mode is generally found. Remember "good amateur practice?" Band plans are a good example.

Band plans vary from region to region, especially on the VHF and higher bands. It's worth taking the time to learn about and follow band plans for the activities, modes and bands that interest you. When everyone works cooperatively and follows the guidelines, it allows each of us to better enjoy the time we spend on the air. Current ARRL Band Plans are at **www.arrl.org/FandES/field/regulations/bandplan.html**. The "Considerate Operators Guide" shown here is a great quick reference to keep handy in your station.

One of the most overlooked areas of good amateur practice is found in §97.313(a): "An amateur station must use the minimum transmitter power necessary to carry out the desired communications." Follow this important rule and you'll be able to conduct your QSO while reducing interference to others on crowded bands.

Many amateurs have an interest in emergency communications and public service. It is one of the core principles of Amateur Radio previously mentioned. It is the responsibility of each licensee to give priority to stations handling emergency communications [§97.101(c)]. Good amateur practice in this area suggests that, in an emergency, you should not transmit unless you can be of direct assistance. While we all wish to be helpful, sometimes the best help in an emergency is to listen, rather than transmit.

Interference

A combination of FCC-mandated and voluntary restrictions are intended to keep us out of one another's way help, but these cannot completely eliminate interference between amateur stations — nor should we expect them to.

Let's put interference into perspective. Note that we're referring only to interference from one amateur station to another, not RFI/TVI or to non-amateur intruders into exclusive ham bands.

Except when it concerns emergency communications, amateur-to-amateur interference is not, in and of itself, illegal.

The Considerate Operator's Frequency Guide

The following frequencies are generally recognized for certain modes or activities (all frequencies are in MHz) during normal conditions. These are not regulations and occasionally a high level of activity, such as during a period of emergency response, DXpedition or contest, may result in stations operating outside these frequency ranges.

Nothing in the rules recognizes a net's, group's or any individual's special privilege to any specific frequency. Section 97.101(b) of the Rules states that "Each station licensee and each control operator must cooperate in selecting transmitting channels and in making the most effective use of the amateur service frequencies. No frequency will be assigned for the exclusive use of any station." No one "owns" a frequency.

It's good practice — and plain old common sense — for any operator, regardless of mode, to check to see if the frequency is in use prior to engaging operation. If you are there first, other operators should make an effort to protect you from interference to the extent possible, given that 100% interference-free operation is an unrealistic expectation in today's congested bands.

Frequencies	Modes/Activities	Frequencies	Modes/Activities
1.800-2.000	CW	14.230	SSTV
1.800-1.810	Digital Modes	14.285	QRP SSB calling frequency
1.810	CW QRP calling frequency	14.286	AM calling frequency
1.843-2.000	SSB, SSTV and other wideband modes	18.100-18.105	RTTY/Data
		18.105-18.110	Automatically controlled data stations
1.910	SSB QRP	18.110	IBP/NCDXF beacons
1.995-2.000	Experimental		
1.999-2.000	Beacons	21.060	QRP CW calling frequency
		21.070-21.110	RTTY/Data
3.500-3.510	CW DX window	21.090-21.100	Automatically controlled data stations
3.560	QRP CW calling frequency	21.150	IBP/NCDXF beacons
3.570-3.600	RTTY/Data	21.340	SSTV
3.585-3.600	Automatically controlled data stations	21.385	QRP SSB calling frequency
3.590	RTTY/Data DX		
3.790-3.800	DX window	24.920-24.925	RTTY/Data
3.845	SSTV	24.925-24.930	Automatically controlled data stations
3.885	AM calling frequency	24.930	IBP/NCDXF beacons
3.985	QRP SSB calling frequency		
		28.060	QRP CW calling frequency
7.030	QRP CW calling frequency	28.070-28.120	RTTY/Data
7.040	RTTY/Data DX	28.120-28.189	Automatically controlled data stations
7.070-7.125	RTTY/Data	28.190-28.225	Beacons
7.100-7.105	Automatically controlled data stations	28.200	IBP/NCDXF beacons
7.171	SSTV	28.385	QRP SSB calling frequency
7.285	QRP SSB calling frequency	28.680	SSTV
7.290	AM calling frequency	29.000-29.200	AM
		29.300-29.510	Satellite downlinks
10.130-10.140	RTTY/Data	29.520-29.580	Repeater inputs
10.140-10.150	Automatically controlled data stations	29.600	FM simplex
		29.620-29.680	Repeater outputs
14.060	QRP CW calling frequency		
14.070-14.095	RTTY/Data		
14.095-14.0995	Automatically controlled data stations		
14.100	IBP/NCDXF beacons		
14.1005-14.112	Automatically controlled data stations		

ARRL band plans for frequencies above 28.300 MHz are shown in *The ARRL Repeater Directory* and on **www.arrl.org**.

Each amateur station has an equal right to operate; just because you've used the same frequency since 1947 doesn't mean you have any more legal right to it than the person who received their license in the mail five minutes ago. The rules specifically prohibit willful or malicious interference [§97.101(d)].

What's malicious interference? Here's an example. If two hams, or groups of hams, find themselves on the same frequency pursuing mutually exclusive objectives, that's happenstance, not malicious interference. On the other hand, if one moves to another frequency and the other follows for the purpose of continuing to cause QRM to the first, the second has crossed the line. If he does it enough, he'll put his license in jeopardy. Of course, what sometimes happens is that they'll all sit on one frequency and argue about who has more right to be there. All it accomplishes is to keep the frequency from being used by anyone for anything worthwhile.

Radio amateurs have the right to pursue legitimate objectives within the privileges conveyed by their licenses, but they also have the obligation to minimize the inconvenience and loss of enjoyment hams cause to others. If there's a tiny segment of a band used for international communication, it's not too much to ask that local rag chews take place elsewhere. If establishing a beacon in the middle of a densely populated area is going to cause interference to nearby weak-signal enthusiasts, an amateur can find another place to put it. And surely, in such cases amateurs don't need the FCC to tell us what growing up in a civilized society should already have taught us to do.

Control Operator Responsibility

Under all circumstances it is the station licensee (or trustee for a club station) who is ultimately responsible for the correct and proper station operation. If the station licensee has designated a control operator, the rules consider both to be equally responsible for the correct operation [§97.103(a)]. This means that the control operator must make certain that the station is being operated properly at all times [§97.105(a)]. If you, as the licensee or control operator, are in doubt as to whether the station is being operated according to the rules and good amateur practice, it is your responsibility to immediately cease the transmissions. Don't transmit again until you are certain the station can resume legal, safe and appropriate operation.

Reciprocal Operating Within the US

Under many circumstances it is legal for licensed amateurs from foreign countries to get on the air while visiting the US. To do so, several criteria must be in order.

1) There must be a current reciprocal operating agreement between the US and their home country. This agreement must be either through a multilateral treaty such as CEPT or IARP (see below) or a direct bilateral agreement between the US and their home country.

2) The foreign licensee must not hold a US amateur license and call sign. If they do, they must operate under the terms of the US license and are not eligible to operate under a reciprocal license agreement.

3) The visitor must not be a US citizen. US citizens must hold a US license to operate and may not operate in the US under any type of reciprocal agreement.

The operating privileges that a visiting amateur may use while in the US depends on several factors. Visitors operating in the US under the European Conference of Postal and Telecommunications Administrations (CEPT) agreement, or under CITEL's International Amateur Radio Permit (IARP) Class 1 agreement, enjoy the full privileges of the Amateur Extra license [§97.301(a)(b)]. Holders of an IARP license other than Class 1 are entitled to VHF and up privileges [§97.301(a)].

Visitors from Canada are entitled to whatever operating privileges they hold at home, but they may not exceed the privileges of the Amateur Extra license [§97.107(a)]. This means that where their Canadian privileges differ from the US allocations (different limits on the phone operating frequencies, for example) they must stay within the privileges for US Amateur Extra licensees.

A visiting amateur who is not from Canada, or is from a country that is not party to CEPT or IARP, may operate if there is an existing bilateral agreement between the US and their home country. They may operate up to the limit of their own privileges back home, again not exceeding Amateur Extra operating privileges.

A visiting amateur from a country with which the US does not hold a reciprocal agreement of any sort may not operate. There is no citizenship requirement for obtaining a US amateur license, though, so any visitor may take the US license exams and then operate with a US call sign. They must provide a permanent US mailing address.

Reciprocal Operating By US Licensees Visiting Other Countries

US licensed amateurs benefit from reciprocal operating agreements when they travel outside the US. As a party to the CEPT agreement, US amateurs visiting most European countries and other countries that are signatories to the CEPT treaty have relatively easy reciprocal operating privileges. In most cases, all you need is to take your US amateur license and a copy of the CEPT agreement (which can be downloaded from **www.arrl.org/FandES/field/regulations/io/cept-ral.pdf**) and you are set to operate. You simply use the appropriate prefix designator before your FCC-issued call sign (for example, DL/N1ND during a visit to Germany).

CITEL's IARP certificate is the basic document necessary for operating in eight Latin/South American countries. IARP does not necessarily grant you instant operating privileges, but it does facilitate the ease of licensing in the countries where it is accepted. To apply for an IARP, download the application from **www.arrl.org/FandES/field/regulations/io/iarp-app.pdf** and submit it, along with the processing fee indicated in the instructions, to the ARRL VEC, 225 Main St, Newington CT 06111. Please allow 2-3 weeks for processing and return mail. Expedited processing is available. Contact the ARRL VEC for more information.

For non-CEPT countries, it is usually necessary for you to make some kind of direct notification or application to the licensing authority in the country you are visiting before you can begin operating. Do not assume that you can simply begin operating even if there is a reciprocal agreement in place. With the cooperation of Veke Komppa, OH2MCN, the ARRL works to maintain a detailed list of requirements for visiting each country in the world. Details are available at **www.arrl.org/**

FandES/field/regulations/io/recip-country.html. If you're planning a trip, it's always a good idea to review the material and start the process early.

Visit **www.arrl.org/FandES/field/regulations/io/faq.html** for the most up-to-date information on reciprocal licensing.

Authorized Transmissions

Normal communications via Amateur Radio are two-way in nature and according to §97.111(a) include:

1) Transmissions necessary to exchange messages with other stations in the amateur service, except those in any country whose administration has notified the ITU that it objects to such communications. The FCC will issue public notices of current arrangements for international communications;

2) Transmissions necessary to meet essential communication needs and to facilitate relief actions;

3) Transmissions necessary to exchange messages with a station in another FCC-regulated service while providing emergency communications;

4) Transmissions necessary to exchange messages with a United States government station, necessary to providing communications in RACES; and

5) Transmissions necessary to exchange messages with a station in a service not regulated by the FCC, but authorized by the FCC to communicate with amateur stations. An amateur station may exchange messages with a participating United States military station during an Armed Forces Day Communications Test.

Practically speaking, you may contact stations in the US and around the world that are authorized to conduct communications with stations in the Amateur Service. Your conversations should be confined to comments of technical or a personal nature when talking to amateurs outside the US [§97.117]. While you may communicate in languages other than English, you are required to identify your station in English if using voice modes. If using CW or digital communications, it is permissible to send your station identification in that mode [§97.119(a)].

Amateurs may engage in a few types of short duration, one-way transmissions spelled out in §97.111(b):

1) Brief transmissions necessary to make adjustments to the station;

2) Brief transmissions necessary to establishing two-way communications with other stations;

3) Telecommand;

4) Transmissions necessary to providing emergency communications;

5) Transmissions necessary to assisting persons learning, or improving proficiency in, the international Morse code;

6) Transmissions necessary to disseminate information bulletins; and

7) Transmissions of telemetry.

Prohibited Communications

While a wide range of communications are legal for Amateur Radio operations, there are also some specific prohibitions. Many of the misunderstandings and disputes among Amateur Radio operators stem from these areas. Although most prohibitions are clear and simple, a few rely on each amateur to decide how to employ good amateur practice. One of the hallmarks of the Amateur Service is the commitment of the licensees to "self-police" our frequencies. A little thought about what we are about to transmit usually keeps us in line.

Certain types of communication are strictly forbidden. These are found in §97.113. For starters, you are never allowed to use the amateur service for business communication. This is defined in §97.113(a)(3) as "Communications in which the station licensee or control operator has a pecuniary interest, including communications on behalf of an employer." The rules do allow on-air swap nets where you can notify other hams of station equipment and accessories for sale — as long as you don't engage in that activity on a regular basis. (The rules don't define "regular basis" so here's a situation where individual amateurs must apply common sense and good amateur practice.)

One of the most frequently asked questions in relation to business communication rules is "Can I, as a paid employee of an emergency response service (such as a hospital, fire department or emergency dispatch office) be 'on the clock' while communicating via Amateur Radio during an emergency?" The FCC clarified this specific area in December 2006, in their Report and Order on Docket 04-140. The Commission's answer is that in their view, in an emergency a responder is being paid for their services as an emergency worker, not as an Amateur Radio operator. Keep in mind that this is specific to emergencies, not routine day-to-day operations.

Another frequently posed question relates to the use of a repeater autopatch to conduct business. The FCC view is that it is permissible to use the autopatch to do such routine tasks as order a pizza or call your doctor's office to tell them you are running late for an appointment. Remember, though, that a repeater owner or trustee may set more stringent standards for the use of their repeater and autopatch than required by the FCC. A good rule of thumb is "if in doubt — don't."

Reciprocal operating agreements and multilateral treaties like CEPT and IARP make it easier for hams to pack up portable stations and get on the air from other countries. *(VK7MO photo)*

Amateurs provide communications free of charge and may not accept compensation for their services [§97.113(a)(2)]. Occasionally hams ask if they may accept T-shirts, hats and the like from organizers of events where hams provide public service communications. The answer is, you cannot be paid directly (money or goods) or indirectly (publicity, advertising, and so on) for your service. If the organizers supply you with T-shirts, hats or other incidental items to identify you while providing your services, that's not considered material compensation. Such items may be accepted, provided the communications would have been provided whether you had received the incidental item or not. This can be one of those "gray areas," so, if in doubt, don't.

Certain transmissions are always prohibited [§97.113(a)(4)]. You may never intentionally transmit music or obscene or indecent words or language on the amateur frequencies. You must not use the amateur frequencies in connection with any criminal activity. You may not transmit information via Amateur Radio in a code or cipher that obscures or hides the meaning of the transmission. False or deceptive messages or station identification are also prohibited. These prohibitions are clear and precise, and the FCC seriously enforces them.

Communications, on a regular basis, that could reasonably be furnished using another radio service are also prohibited [§97.113(a)(5)]. For example, you might help out your local firefighters during an emergency, but they need to rely on authorized public safety frequencies and radios for their day-to-day operations.

Broadcasting

Can you broadcast traffic reports and similar information on the local repeater? The answer is a resounding *no*! Part 97.113(b) is clear: Broadcasting is not permitted on the amateur bands. The FCC defines broadcasting as "transmissions intended for reception by the general public, either direct or relayed" [§97.3(a)(10)]. You may think it is helpful as a public service to relay or retransmit traffic reports or newscasts on the local repeater, but Amateur Radio transmissions are not intended to be transmitted to and received by the general public. Even allowed one-way amateur transmissions, such as code practice or HF propagation bulletins, are intended for an audience of licensed Amateur Radio operators, not the general public.

Along these lines, you are not allowed to retransmit commercial AM or FM radio broadcasts or television audio or video. The two allowed specific retransmissions are weather bulletins transmitted by US government stations (such severe weather alerts from the various NOAA weather radio stations) and, with permission, NASA manned space communications. In both cases, remember that you may not conduct such transmissions on a routine basis.

Amateur Radio's reputation for providing accurate information means that amateur transmissions are occasionally used as a source of information by local television and radio stations. Broadcasters are aware of SKYWARN and other emergency activities and will occasionally use information from Amateur Radio communications to help with news programming. There is a fine line between what is acceptable and what is not in this area.

In an emergency, Amateur Radio can be used to directly provide news information if several conditions are met:

1) The information must be directly related to the event;
2) It must involve the immediate safety of life or property; and
3) No other means of communications is available at the time of the event.

The general rule is that Amateur Radio is not to be used for newsgathering for production purposes by the media [§97.113(b)]. This is an important protection intended to stop encroachment by commercial news media who may see Amateur Radio as an inexpensive alternative to other communications systems.

Remember as well that the FCC rules provide some flexibility to allow the amateur community to meet its basic purpose of providing emergency communications. In a disaster or emergency, it is your responsibility as a licensed amateur to provide whatever assistance you can with emergency communications.

Third Party Communications

In many cases it is permissible for a licensed amateur to provide communications on behalf of someone other than the control operator or station licensee. This is known as *third party communications* and is defined in §97.3(a)(46) as "A message from the control operator (first party) of an amateur station to another amateur station control operator (second party) on behalf of another person (third party)."

Third party communications usually fall into one of three main types:

1) *Third party messages* — written messages generally sent via traffic nets or Amateur Radio message-handling services (packet or radio e-mail);
2) *Telephone interconnection* — autopatch or phone patch communications;
3) *Direct participation* — where the third party actively participates in transmitting the message.

FCC licensed amateurs may conduct third party traffic if specific conditions are met [§97.115]. A US amateur station may transmit third-party communications to another amateur station in the US. Third party traffic is permitted with stations in foreign countries for the purpose of emergency or disaster relief communications. For routine (non-emergency) third party communications, though, there must be an agreement between the US and the government of the country where the other amateur station is located. A current list of countries with which the US has third party agreements is found in **Table 3**.

The control operator must always be present at the control point when third party communications are being conducted, and the control op must continuously monitor and supervise the transmissions. Third party communications may not be conducted on behalf of anyone whose US license has been revoked; is currently suspended; or has been surrendered in lieu of revocation, suspension or monetary forfeiture. Nor may third-party communications be conducted on behalf of anyone subject to a cease-and-desist-order relating to Amateur Radio operations [§97.115(b)(2)]. Also, no station may transmit third party communications while under automatic control unless that station is using RTTY or data emissions [§97.115(c)].

Station Identification

There is a simple reason why stations must transmit their call sign — so people will know who they are talking to. In addition, unidentified transmissions are prohibited. The rules are straightforward in this area. Part 97.119(a) states: "Each amateur station, except a space station or telecommand station, must transmit its assigned call sign on its transmitting channel at the end of each communication, and at least every 10 minutes during a communication, for the purpose of clearly making the

Table 3
Third-Party Traffic Agreements List

Countries that share third-party traffic agreements with the US:

Occasionally, DX stations may ask you to pass a third-party message to a friend or relative in the States. This is all right as long as the US has signed an official third-party traffic agreement with that particular country, or the third party is a licensed amateur. The traffic must be noncommercial and of a personal, unimportant nature. During an emergency, the US State Department will often work out a special temporary agreement with the country involved. But in normal times, never handle traffic without first making sure it is legally permitted.

US Amateurs May Handle Third-Party Traffic With:

C5	The Gambia	XE	Mexico
CE	Chile	YN	Nicaragua
CO	Cuba	YS	El Salvador
CP	Bolivia	YV	Venezuela
CX	Uruguay	ZP	Paraguay
D6	Federal Islamic Rep. of the Comoros	ZS	South Africa
DU	Philippines	3DA	Swaziland
EL	Liberia	4U1ITU	ITU - Geneva
GB	United Kingdom	4U1VIC	VIC - Vienna
HC	Ecuador	4X	Israel
HH	Haiti	6Y	Jamaica
HI	Dominican Republic	8R	Guyana
HK	Colombia	9G	Ghana
HP	Panama	9L	Sierra Leone
HR	Honduras	9Y	Trinidad and Tobago
J3	Grenada		
J6	St Lucia		
J7	Dominica		
J8	St Vincent and the Grenadines		
JY	Jordan		
LU	Argentina		
OA	Peru		
PY	Brazil		
TA	Turkey		
TG	Guatemala		
TI	Costa Rica		
T9	Bosnia-Herzegovina		
V2	Antigua and Barbuda		
V3	Belize		
V4	St Kitts and Nevis		
V6	Federated States of Micronesia		
V7	Marshall Islands		
VE	Canada		
VK	Australia		
VP6	Pitcairn Island*		

Notes:

*Since 1970, there has been an informal agreement between the United Kingdom and the US, permitting Pitcairn and US amateurs to exchange messages concerning medical emergencies, urgent need for equipment or supplies, and private or personal matters of island residents.

Please note that Region 2 of the International Amateur Radio Union (IARU) has recommended that international traffic on the 20 and 15-meter bands be conducted on 14.100-14.150, 14.250-14.350, 21.150 -21.200 and 21.360-21.450 MHz. The IARU is the alliance of Amateur Radio societies from around the world; Region 2 comprises member-societies in North, South and Central America and the Caribbean.

At the end of an exchange of third-party traffic with a station located in a foreign country, an FCC-licensed amateur must transmit the call sign of the foreign station as well as his own call sign.

Current as of January 2007; see www.arrl.org/FandES/field/regulations/io/3rdparty.html for the latest information.

---

source of the transmissions from the station known to those receiving the transmissions. No station may transmit unidentified communications or signals, or transmit as the station call sign, any call sign not authorized to the station."

You will hear a wide range of comments and opinions and variations on this simple rule. But under FCC rules you are required to give your call sign every 10 minutes during active communications and at the end of the contact you are finishing. You are not *required* to give your call sign at the start of a contact. (Common sense suggests that during routine operation you would probably want to send your call sign at the start of the contact so that the other station knows who is talking to them.) You do not have to ID your station in a roundtable discussion every 10 minutes if you haven't transmitted since the last time you identified.

A few additional rules apply to station identification requirements:

1) You should transmit your station ID using the mode in which you are communicating.

2) When conducting an international third party contact, you must give both the call sign of the station with which you are communicating and your own call sign.

3) If you are transmitting the station ID of an automatically controlled station using CW, the speed may not exceed 20 WPM.

4) If you have upgraded your license recently, you must sign the correct temporary designator, such as "temporary AG" or "temporary AE" until your upgrade is processed by the FCC and appears in the ULS database.

SUBPART C — SPECIAL OPERATIONS

As your experience grows and as your interests change, you may want to become involved in more specialized activities. These are not different "modes" of operation — they are different "uses" for Amateur Radio stations. When you are talking on or controlling a local repeater, sending or receiving messages through a message forwarding system such as radio e-mail or packet, or communicating through one of the Amateur Radio satellites, you are participating in a special operation, and special rules apply to your activities.

Internet Assisted Amateur Radio

By Brennan Price, N4QX and ARRL General Counsel Chris Imlay, W3KD

Internet Assisted Amateur Radio (also called Voice Over the Internet Protocol or VoIP) has had a major effect on the transfer of information and Amateur Radio is no exception. The name of the protocol varies, depending on the system being used. It might be called EchoLink, IRLP (Internet Repeater Linking Project) or some other means, but it is still a combination of Amateur Radio transmissions and an interface to the Internet. Here is a short Q&A format addressing the legalities of their usage:

What Part 97 regulations govern VoIP-assisted Amateur Radio?

In short, all of them or none of them, depending on whether you're asking about the "VoIP-assisted" or the "Amateur Radio" porting of VoIP-assisted Amateur Radio.

Many people focus on the novelty of the Internet when asking questions about the legal uses of certain systems. Such focus is misdirected. Part 97 doesn't regulate *systems*; it regulates *stations*. The FCC cares that the station – not the Internet but the Amateur Radio station – is properly operated. And all of the rules that apply to any Amateur Radio station apply to one that retransmits audio fed to it by VoIP.

Since the FCC doesn't care about the Internet part, are there any particular rules of which a ham considering such an operation should be aware?

The obvious answer is all of them, but the focus is on a few that are easy to overlook, particularly for stand-alone, single channel operations. The main points to remember are:

- All stations must be controlled
- Only certain types of stations may be automatically controlled
- Simplex voice operations do not qualify for automatic control
- Any station that is remotely controlled via radio must utilize an auxiliary station to execute said control and the auxiliary station must operate on frequencies where they are permissible (97.201)

It is not as hard as it sounds. All you have to do is think about the type of station you are operating and how it is controlled. Let's look at a few examples:

Two automatically controlled repeaters are linked via VoIP. Is this legal?

Absolutely. Forget the VoIP link, because that is the Internet. We are talking about two repeaters. Repeaters are legal. Repeaters may be automatically controlled. And repeaters may be linked. This is no difference between this set-up and two repeaters being linked by another wired mechanism or by auxiliary station. The only caveat is that the VoIP software must exclude nonhams from accessing the repeaters from the Internet. The key is to avoid any configuration that would (1) permit a nonham to key an amateur transmitter without the presence of a control operator; and (20 prevent the initiation by a nonham of a message via Amateur Radio without the presence of a control operator.

Is it permitted to enable automatically controlled simplex nodes?

No. Only certain types of Amateur Radio stations may be operated unattended, under automatic control. These are space stations, repeaters, beacons auxiliary stations and certain types of stations transmitting RTTY or data emissions. Simplex VoIP nodes are neither none of these. Therefore, none of the stations that qualify for automatic control describe a VoIP node, so such a station must be locally or remotely controlled (as any Amateur Radio station is allowed to be).

Locally or Remotely Controlled – what does that mean?

A simplex VoIP node may be locally controlled by an operator present at the node, which is the control point. A VoIP node may be remotely controlled at some other point, with the operator issuing commands via a wireline or radio-control link. If a radio control link is used, it must use an auxiliary station and operate on a frequency on which auxiliary stations are permitted. It is this *remotely-controlled* aspect that allows VoIP nodes – as long as they ar eon the right band.

Let's consider some options:

- A control operator is stationed an active at the VoIP node on any frequency. This is a locally controlled station, not unlike a typical FM simplex operation. ***This is legal***
- A control operator communicates with and controls a simplex VoIP node with a handheld, transmitting and listening to the node on 223.52 MHz. This is wireless remote control and operates on a frequency where an auxiliary station is permitted. ***This is legal.***
- A control operator communcations with and control a VoIP node with a handheld, transmitting on the node frequency of 145.85 MHz.. A wireless remote control must utilize an auxiliary station and 145.85 MHz is not an authorized auxiliary station frequency. ***This operation is not legal***. It may be made legal by either locally controlling the VoIP node or controlling it by an auxiliary station operating on a permissible frequency.
- A control operator is using a VoIP node on 145.85 MHz but is controlling the node via a dedicated internet connection or dedicated telephone line. ***This is legal.*** The VoIP node is being controlled via a wireline connection
- A control operator is usingd dual-band HT and operating a simplex VoIP node on 145.85 MHz but is sending control commands to the node using 223.52 MHz while continuously monitoring the VoIP node. ***This is legal.*** The VoIP node is being controlled by an auxiliary station on an appropriate frequency and may be controlled in this manner.
- Same configuration as above except the control operator does not continuously monitor the VoIP node's transmission. ***This operation is not legal.*** When a simplex VoIP node is enabled, it must be continually attended, either locally or remotely. A simplex VoIP node is no different than other FM simplex operations and such operations may not be automatically controlled.

Auxiliary Stations

Amateur Radio stations are sometimes set up and controlled remotely via an RF link. A station involved in this type of activity is defined under §97.3(a)(7) as an *auxiliary station*. Auxiliary stations are often set to perform such tasks as controlling or linking repeaters or performing station control tasks, such as changing operating parameters of a remote base station. The stations that these auxiliary stations control, as well as the auxiliary stations themselves, become what the FCC refers to as a "system of cooperating stations."

Auxiliary stations have some specific operating guidelines that enable them to carry out their function while holding to the principle of effective use of our Amateur Radio spectrum. They may operate only on frequencies in the 2-meter band or higher, with the exception of these specific segments: 144.0-144.5, 145.8-146.0, 219-220, 222.00-222.15, 431-433 and 435-438 MHz [§97.201(b)]. Auxiliary stations may be automatically controlled and may transmit one-way communication [§97.201(d)(e)]. Holders of a Technician or higher license may be an auxiliary station or serve as control operator for such a system [§97.201(a)]. Finally, in cases where there is interference between two auxiliary stations, both are equally responsible to resolve the issue, unless one of the stations has been coordinated by a frequency coordinating body. In that case, the non-coordinated auxiliary station has primary responsibility to resolve the issue [§97.201(c)].

Remote Operation and Links

In today's time of deed and antenna restrictions, many amateurs find it convenient or even necessary to employ a *remote base station*. While not defined in Part 97, a remote base uses a form of auxiliary operation to both control and use the station. As such, they must follow the rules for auxiliary operation as well as those for remote control of a station. Remote control is defined in §97.3(a)(38) as "The use of a control operator who indirectly manipulates the operating adjustments in the station through a control link to achieve compliance with the FCC Rules."

Remote bases are not the same as repeaters. In a remote base system, the user is part of the system, whereas in a repeater system, a user is not part of the system. This means that the person transmitting on a remote base must be a control operator of the system, or working directly under the supervision of a control operator. This is different from a repeater where individual users do not have to be able to control the operation of the repeater itself.

With the advent of the modern multiband VHF/UHF radios, it has become common for individuals to use those radios for "crossband repeating." Most often, operation as a crossband repeater is actually operation as a remote base, and as such must meet the remote base rules.

Until the rules changed in December 2006, auxiliary stations were required to operate on frequencies above 222.15 MHz. It is now legal to operate an auxiliary station on the 2-meter band with the exception of 144.0-144.5 MHz and 145.8-146.0 MHz. This makes it easier to legally operate these multiband radios as remote bases. They must still have some sort of system for a control link (remember, all stations must be controlled by some legal mechanism). They must also have some sort of timer or means of shutting down the system after three minutes should the control link fail [§97.213]. But with auxiliary stations now allowed on 2 meters, the user's ID over both sides of the link serves to legally ID the remote station when operating as a crossband repeater.

It is also common to link repeaters to one another or to an Internet connection (such as IRLP, EchoLink or simply a computer gateway). This linking of radio constitutes a system of cooperating stations and involves the concepts of remote bases and auxiliary stations. The auxiliary station rules apply. One of the big concerns to keep in mind when participating in interconnection to the Internet is to ensure that there is no access to Amateur Radio via the Internet by a non-amateur without an appropriate control operator. As with all amateur communications, the Part 97 rules for station identification, interference and other normal operating issues apply.

Repeater Stations

Almost every ham uses repeaters at one time or another. A repeater is defined in §97(a)(39) as "An amateur station that simultaneously retransmits the transmission of another amateur station on a different channel or channels." Any station owned by a Technician or higher licensee may operate as a repeater on 6 meters and up, and General or higher licensees may also operate as a repeater on 10 meters. Repeater operation is not allowed in these segments of those bands: 28.0-29.5, 50.0-51.0, 144.0-144.5, 145.5-146.0, 222.0-222.15, 431.0-433.0 and 435.0-438.0 MHz [§97.205(b)].

By their nature, repeaters tend to be placed in locations that are not easily accessible, such as mountaintops or tall buildings.

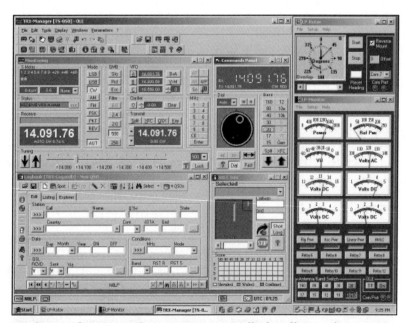

Modern software and computer-controlled radios make remote operation easier than ever, but you need to follow the rules. Be sure to review the requirements in Subpart C for details on auxiliary and remote base operation.

This means they are frequently controlled remotely. Remote control can be achieved through several means. A dedicated wireline or non-published telephone line is permissible. If the repeater also has an autopatch interface to the telephone line, there must be an additional control link. That's needed because it would not be possible to terminate operation of the repeater via the telephone line if the problem was being caused by the failure of that line. You may also use a radio control link through an auxiliary station as a means of remote control for the repeater station. It is also permissible for a repeater to be under automatic control [§97.205(d)]. It is important to remember that regardless of the type of control, the licensee is *always* responsible for the proper operation of the repeater.

One of the most common repeater problems involves repeater users who transmit communications that violate the rules — obscene material or music, for example. While the rules state that the control operator is not accountable when such actions occur [§97.205(g)], it is still their responsibility to ensure that the repeater is being operated under the rules. In such cases the control operator should shut down the repeater until the problem is resolved. If necessary to ensure proper operation of the repeater, it is permissible for the repeater owner to limit its use to only certain stations [§97.205(e)]. A repeater is not a "public utility." No licensee has the automatic right to use a specific repeater. Repeater owners may always make rules for use of their equipment that are more stringent than those imposed by Part 97. Failure to abide by those additional rules can lead to the revocation of your privilege to use that specific repeater.

You won't have to spend much time on the VHF and UHF bands to realize that they are crowded. To deal with this congestion, a series of frequency coordinating groups have developed to try and manage the repeater spectrum. From §97.3(A)(22): A frequency coordinator is "An entity, recognized in a local or regional area by amateur operators whose stations are eligible to be auxiliary or repeater stations, that recommends transmit/receive channels and associated operating and technical parameters for such stations in order to avoid or minimize potential interference."

These coordinating groups work on local or regional policies to maximize the efficient use of the available spectrum as well as minimize interference issues or problems. The National Frequency Coordinator's Council (NFCC) certifies most frequency coordinating groups. Each group develops procedures for a designated territory and maintains an accurate database of coordinated repeaters within that area.

When repeater owners experience harmful interference, coordinators assist in resolving the problems. If an interference issue involves a coordinated and uncoordinated repeater, §97.205(c) states that the primary responsibility for resolving the issue rests with the uncoordinated repeater.

Message Forwarding Systems

Improvements in technology mean changes in our ability to communicate with new methods. As with any operating method — new or old — we need to ensure that we adhere to the rules. Part 97.3(a)(31) defines a message forwarding system as "A group of amateur stations participating in a voluntary, cooperative, interactive arrangement where communications are sent from the control operator of an originating station to the control operator of one or more destination stations by one or more forwarding stations." These systems now operate on HF, VHF and UHF and often a message may pass among several frequency bands as it makes its way to the recipient.

The licensee of the originating station of a message is always primarily accountable for any message content that violates FCC rules [§97.219(b)]. But the responsibility does not stop there. The first forwarding station must make sure that the originator of a message is authorized to send messages over the system and must take responsibility for any messages it retransmits into a system. In essence it has the responsibility to act as a "filter" to protect other stations down the line [§97.219(d)(1),(2)]. However subsequent forwarding stations also have the responsibility to terminate any message they are aware of that is in violation of the rules, even though they are not responsible for it reaching them [§97.219(3)].

Remember that Part 97 rules apply at all times. The rules regarding third-party traffic, business communications, obscene and indecent materials, and so forth apply to messages sent via message forwarding systems just as they do to "real time" amateur communications.

Two or more unattended HF digital stations that are connected may be automatically controlled while transmitting RTTY or data emissions, but they are restricted to specific frequencies. There are no restrictions at 6 meters and above. On HF, only these segments are allowed: 28.120-28.189, 24.925-24.930, 21.090-21.100, 18.105-18.110, 14.0950-14.0995, 14.1005-14.112, 10.140-10.150, 7.100-7.105 and 3.585-3.600 MHz [§97.221(b)].

Automatic control is authorized only if the station is responding to an interrogation from a station under local or remote control. In addition, no transmission from an automatically controlled system may occupy a bandwidth of more than 500 Hz [§97.221(c)(1),(2)].

An automatically controlled station may still be placed on a frequency outside of the allowed fully automated sub-bands as long as a manually controlled station (a real person) initiates or "turns on" the station. Until that happens, it must remain silent.

SUBPART D — TECHNICAL STANDARDS

Many amateurs find this subpart of the rules to be the most confusing — or least interesting. It's an important section because here you will find the specifics rules that determine the frequencies, transmitter power and emission types you may use. This subpart also governs the quality of the signals you are allowed to transmit.

License Classes

Currently there are three levels of Amateur Radio license issued in the US — Technician, General and Amateur Extra. Each level of license requires you to pass an examination administered by a team of at least three Volunteer Examiners (VEs) working under the direction of an FCC-certified Volunteer Examiner Coordinator (VEC).

The Technician license requires that you correctly answer a minimum of 26 questions on a 35-question written test. To earn the General license, you pass the Technician exam and then an additional 35-question General exam (again with 26 correct answers minimum). The Amateur Extra license requires that you pass the Technician and General examinations, as well as a 50-question Amateur Extra exam (with at least 37 correct answers).

The written exam questions cover rules and regulations, radio and electronic theory, antennas, safety, operating techniques and other amateur practices. The FCC no longer requires candidates to pass a Morse code examination for any amateur license.

There are two additional US license classes — Novice and Advanced. No new Novice or Advanced licenses have been issued since April 15, 2000. Current Novice and Advanced licensees may continue to renew at expiration and use frequencies, modes and power levels for that license as allowed in the rules.

Frequency Bands

Bands of frequencies are allocated to the Amateur Service from 1800 kHz (73 kHz in the UK) to over 300 GHz. **Table 4** gives an overview of the radio spectrum and shows the Amateur Service bands allocated in the International Telecommunication Union (ITU) Radio Regulations.

Development of band plans is an ongoing process. It requires planners to research, invite and digest comment from amateurs, arrive at a mix that will serve the diverse needs of the amateur community, and adopt a formal band plan. This is a process that can take a year or more on the national level and a similar period in the International Amateur Radio Union (IARU). Nevertheless, new communication modes or changes in the popularity of existing ones can make a year-or-two-old band plan look obsolete.

Such revolutionary change has taken place in the past decade with the popularity of new digital modes. Changes of this magnitude cause the new users to scramble for frequen-

Table 4
The Electromagnetic Spectrum with Amateur Service Frequency Bands by ITU Region

Wave-length	Frequency	Nomen-clature	Metric Band	Amateur Radio Bands by ITU Region		
				Region 1	Region 2	Region 3
1 mm	300 GHz	EHF Milli-metric	1 mm	241-250	241-250	241-250
			2 mm	142-149	142-149	142-149
			2.5 mm	119.98-120.02	119.98-120.02	119.98-120.02
			4 mm	75.5-81	75.5-81	75.5-81
			6 mm	47-47.2	47-47.2	47-47.2
1 cm	30 GHz	SHF Centi-metric	1.2 cm	24-24.25	24-24.25	24-24.25
			3 cm	10-10.5	10-10.5	10-10.5
			5 cm	5.65-5.85	5.65-5.925	5.65-5.85
			9 cm		3.3-3.5	3.3-3.5
10 cm	3 GHz	UHF Deci-metric	13 cm	2.3-2.45	2.3-2.45	2.3-2.45
			23 cm	1240-1300	1240-1300	1240-1300
			33 cm		902-928	
			70 cm	430-440	430-440	430-440
1	300 MHz	VHF Metric	1.25 m		222-225	
			2 m	144-148	144-148	144-148
			6 m		50-54	50-54
10	30 MHz	HF Deca-metric	10 m	28-29.7	28-29.7	28-29.7
			12 m	24.89-24.99	24.89-24.99	24.89-24.99
			15 m	21-21.45	21-21.45	21-21.45
			17 m	18.068-18.168	18.068-18.168	18.068-18.168
			20 m	14-14.350	14-14.350	14-14.350
			30 m	10.0-10.150	10.1-10.150	10.1-10.150
			40 m	7-7.1	7-7.3	7-7.1
			80 m	3.5-3.8	3.5-4	3.5-3.9
100	3 MHz	MF Hectometric	160 m	1.81-1.85	1.8-2	1.8-2
1000	300 kHz	LF Kilometric				
10,000	30 kHz	VLF Myriametric				
100,000	3 kHz					

Nomenclature bracket spans: EHF through SHF grouped as "Microwaves".

Note: This table should be used only for a general overview of where Amateur Service and Amateur-Satellite Service frequencies by ITU Region fall within the radio spectrum. They do not necessarily agree with FCC allocations; for example, the 70-cm band is 420-450 MHz in the United States.

Subpart D is the Technical Standards section. This subpart includes rules governing FCC Certification of certain types of Amateur Radio equipment, including power amplifiers.

cies and some of the existing mode users to draw their wagons in a circle. The national societies (such as ARRL), their staffs and committees, and the IARU have the job of sorting out the contention for various frequencies and preparing new band plans. Fortunately, we have not exhausted all possible ways of improving our management of the spectrum so that all Amateur Radio interests can be accommodated.

Here are some band-by-band highlights:

The 160-Meter Band

The 160-meter band, 1800-2000 kHz, provides some excellent DX opportunities in addition to local operations. The basic problem with allocations in this band has been competition with the Radiolocation Service. New pressures are possible as a result of planned expansion of the medium-frequency broadcast band in the 1605-1705 kHz range.

The 80-Meter Band

While US amateurs enjoy the use of 3500-4000 kHz, not all countries allocate such a wide range of frequencies to the 75/80-meter band. There are fixed, mobile and broadcast operations, particularly in the upper part of the band.

The 60-Meter Band

The FCC has granted amateur *secondary* access on upper sideband (USB) *only* to five discrete 2.8-kHz wide channels at the following frequencies:

Channel Center	Amateur Tuning Frequency
5332 kHz	5330.5 kHz
5348 kHz	5346.5 kHz
5368 kHz	5366.5 kHz
5373 kHz	5371.5 kHz
5405 kHz	5403.5 kHz

While the center channel is the allocated frequency, on USB, amateurs must set their transceivers to the amateur tuning frequency. *This is very important.* General, Advanced and Amateur Extra licensees may operate on these channels with no more than 50 W PEP ERP (effective radiated power). In this case, ERP is calculated by multiplying transmitter power by antenna gain relative to a dipole. That means 50 W to a dipole is the maximum allowed, but if you use an antenna with more gain than a dipole you must reduce transmitter power accordingly. If you use an antenna other than a dipole, you must include information about its gain characteristics in your station log [§97.303(s)].

The 40-Meter Band

The 40-meter band has a big problem: International broadcasting occupies the 7100-7300 kHz band in many parts of the world. That changes on March 29, 2009, when the 7100-7200 kHz band will become amateur exclusive. International broadcast stations will no longer be allowed in this segment — clearer frequencies on this band are coming.

During the daytime, particularly when sunspots are high, broadcasting does not cause much interference to US amateurs. At night, however, especially when sunspot activity is low, the broadcast interference is heavy. Some countries allocate only the 7000-7100 kHz band to amateurs. Others, particularly in Region 2, allocate 7100-7300 kHz as well, which at times is subject to interference. The result is that there is a great demand for frequencies in the 7000-7100 kHz segment. The effect is that there are two band plans overlaid on each other: ours, spread out over 7000-7300 kHz and another one that compresses everything into 7000-7100 kHz. The pending changes in 2009 will help resolve some of this compression, giving the amateur community worldwide more spectrum.

The 30-Meter Band

The 30-meter band, 10100-10150 kHz, is excellent for CW and digital modes. The only problem is that US amateurs must not cause harmful interference to the fixed operations outside the US. This restricts transmitter power output to 200 W and is one reason for not having contests on this band.

The 20-Meter Band

The workhorse of DX is undoubtedly the 20-meter band, 14000-14350 kHz. It offers excellent propagation to all parts of the world throughout the sunspot cycle and is virtually clean of interference from other services.

The 17-Meter Band

The 18068-18168 kHz band was awarded to amateurs on an exclusive basis, worldwide, at WARC-79. It was made available for amateur use in the US in January 1989. It shares propagation characteristics with 15 and 20 meters.

The 15 and 12-Meter Bands

The 21000-21450 and 24890-24990 kHz bands are excellent for DX during the high part of the sunspot cycle. They also offer some openings throughout the rest of the sunspot cycle.

The 10-Meter Band

Spanning 28000-29700 kHz, this is an exclusive amateur band worldwide. Its popularity rises and falls with sunspot numbers and propagation.

The VHF and Higher Bands

The 6-meter band is not universal, but the trend seems to be toward allocating it to amateurs as TV broadcasting vacates the 50-54 MHz band. It is also excellent for amateur exploitation of meteor-scatter communications using various modes including digital.

Two meters is heavily used throughout the world for CW, EME (moonbounce), SSB, FM and packet radio. The allocation is 144-148 MHz. Satellites occupy the 145.8-146 MHz segment.

US amateurs have a primary allocation at 222-225 MHz, which is largely used for repeaters. The Commission has allocated 219-220 MHz to the Amateur Radio Service on a secondary basis, only for stations participating in fixed, point-to-point digital messaging systems. There are special provisions to protect domestic waterways telephone systems using that band.

The 70-cm band is prime UHF spectrum. The 430-440 MHz band is virtually worldwide, whereas the 420-430 MHz and 440-450 MHz bands are not. Frequencies around 432 are used for weak-signal work, including EME, and the 435-438 MHz band is for amateur satellites. The 70-cm band is the lowest frequency band that can be used for fast-scan television and spread spectrum emissions.

The 33-cm band (902-928 MHz) is widely shared with other services, including Location and Monitoring Service (LMS), which is primary, and ISM (industrial, scientific and medical) equipment applications. A number of low-power devices including spread spectrum local area networks operate in this band under Part 15 of the FCC's Rules.

The 1240-1300 MHz band is used by Amateur Radio operators for essentially all modes, including FM and packet. By regulation, the 1260-1270 MHz segment may be used only in the Earth-to-space direction when communicating with amateur satellites.

Amateur Radio is primary at 2390-2417 MHz. While the Amateur Service has a secondary allocation in the 2300-2450 MHz band in the international tables, in the United States the allocation is 2390-2400 MHz and 2390-2417 MHz primary, and 2300-2310, and 2417-2450 secondary. Most of the weak-signal work in the US takes place around 2304 MHz, while much of the satellite activity is in the 2400-2402 MHz segment.

The remaining microwave and millimeter bands are the territory of amateur experimenters. It is important that the Amateur Service and the Amateur Satellite Service use these bands, and contribute to the state-of-the-art in order to retain them. There is growing interest on the part of the telecommunications industry and the space science community to fully exploit the 20-95 GHz spectrum.

It is important to remember that we share frequency allocations on many of our bands above 420 MHz. For example, while we have a frequency allocation between 420 and 450 MHz, we are only the secondary user (with the Radiolocation Service designated as primary). In any case where an amateur station as a secondary user causes harmful interference to the primary user (regardless of band), it is the sole responsibility to mitigate or eliminate that interference.

Frequency Allocations and Emission Types

Part 97.301 lays out the specific frequency allocations for each type of amateur license, and §97.305 delineates the types of emissions that are permissible on each portion of each band. The chart in **Fig 2** summarizes this information.

Emission Standards

In keeping with the principle of good amateur practice, it is important that the signals transmitted by an amateur station be "clean." But it is for more than on-the-air aesthetics — good signal quality reduces interference and problems on the bands, which in turn makes our operating time more enjoyable and easier.

The rules are clear about signal quality. No station should occupy more bandwidth than is necessary for the type of communication being conducted [§97.307(a)]. Your modulated signal must not exceed the band segments authorized for your license [§97.307(b)]. Spurious emissions must be reduced as much as possible and corrected if they are causing harmful interference [§97.307(c)]. The remaining paragraphs of 97.307 deal with specifications for various modes.

Certification and Standards for External RF Power Amplifiers

Certain types of Amateur Radio equipment are required to have FCC Certification (formerly known as type-acceptance). Primarily this is done to combat the modification of Amateur Radio gear for illegal use on frequencies in and around the Citizen's Band Radio Service frequencies (commonly referred to as 11 meters). New RF power amplifiers are required to exhibit no amplification between 26 and 28 MHz, and they may not be designed to allow easy modification to do so [§97.317(a)(3) and (b)].

SUBPART E — PROVIDING EMERGENCY COMMUNICATIONS

Providing emergency communications is one of the basic purposes of the Amateur Radio service. Because of its importance, Part 97 devotes a small but significant Subpart to special rules for use during emergencies.

Emergency communication means "essential communication needs in connection with the immediate safety of human life and immediate protection of property when normal communication systems are not available" [§97.403(b)]. This rule section in essence states when life or property are at risk, nothing in the Part 97 rules prevents the licensee from using Amateur Radio to try to obtain the help or relief needed. Part 97.405 goes further when it states that a station in distress should use *any means at its disposal* to "attract attention, make known its condition and location, and obtain assistance." As FCC officials have stated, "In a real emergency — do what is necessary."

This isn't a responsibility to be taken lightly. The FCC records are full of enforcement actions involving stations that "cried wolf" or made false claims. The trust that the FCC places in the amateur community gives us a lot of latitude during emergencies, but it is a trust that each licensee must responsibly bear.

Often when participating in an emergency (or when practicing for them through drills, nets or public service events) you may use self-assigned "tactical" call signs during your communications. While these are not prohibited by Part 97, remember that you are still required to give your FCC-assigned

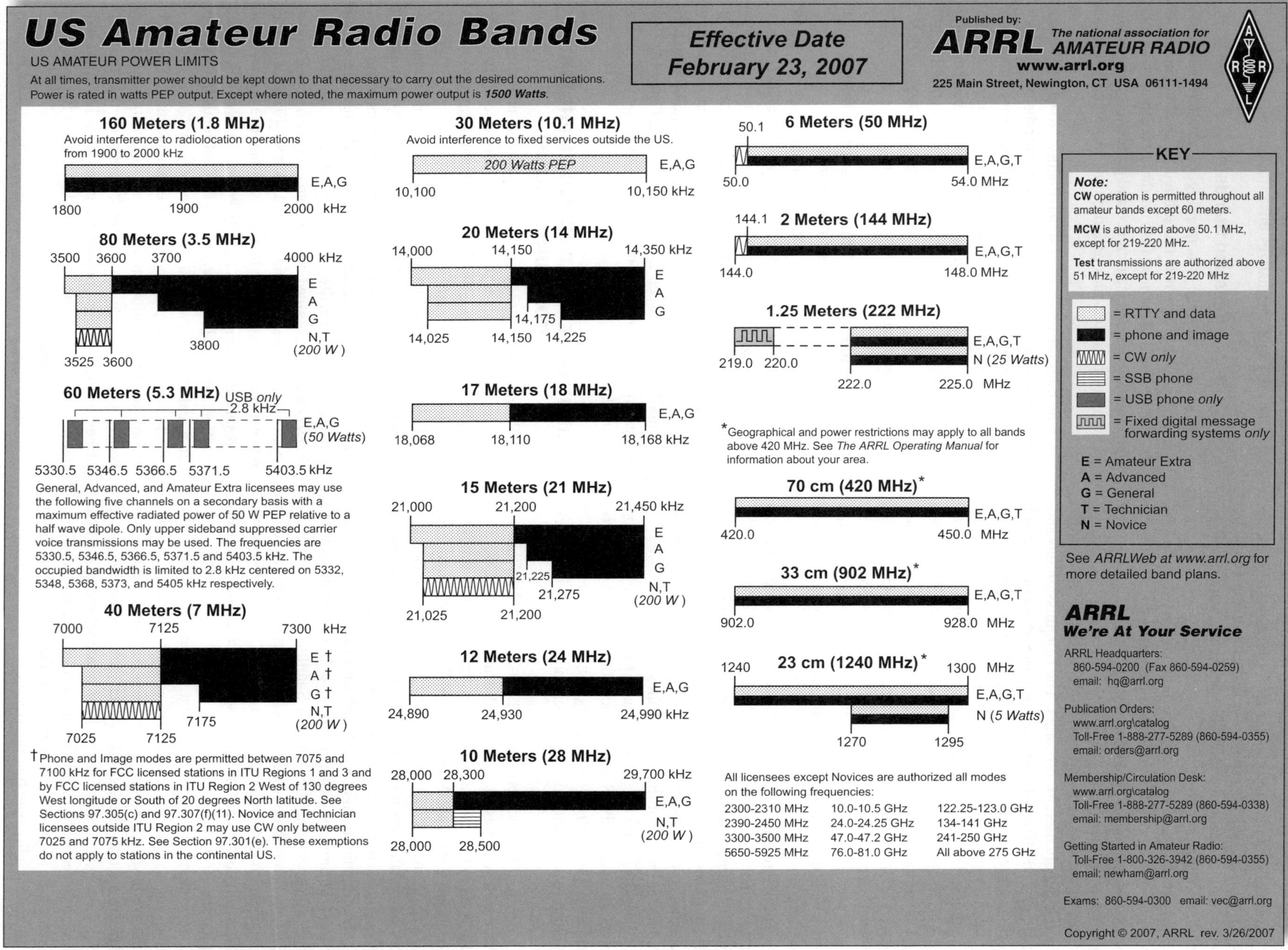

US Amateur Radio Bands

Published by: ARRL — The national association for AMATEUR RADIO
225 Main Street, Newington, CT USA 06111-1494
www.arrl.org

Effective Date February 23, 2007

US AMATEUR POWER LIMITS

At all times, transmitter power should be kept down to that necessary to carry out the desired communications. Power is rated in watts PEP output. Except where noted, the maximum power output is **1500 Watts**.

KEY

Note: CW operation is permitted throughout all amateur bands except 60 meters.
MCW is authorized above 50.1 MHz, except for 219-220 MHz.
Test transmissions are authorized above 51 MHz, except for 219-220 MHz.

- ▦ = RTTY and data
- ■ = phone and image
- ▨ = CW only
- ▥ = SSB phone
- ▤ = USB phone only
- ▦ = Fixed digital message forwarding systems only

- E = Amateur Extra
- A = Advanced
- G = General
- T = Technician
- N = Novice

See *ARRLWeb* at *www.arrl.org* for more detailed band plans.

160 Meters (1.8 MHz)
Avoid interference to radiolocation operations from 1900 to 2000 kHz.
1800 — 1900 — 2000 kHz — E,A,G

80 Meters (3.5 MHz)
3500 — 3600 — 3700 — 3800 — 4000 kHz
3525 — 3600 — N,T (200 W)
E / A / G

60 Meters (5.3 MHz)
USB only, 2.8 kHz
5330.5, 5346.5, 5366.5, 5371.5, 5403.5 kHz — E,A,G (50 Watts)

General, Advanced, and Amateur Extra licensees may use the following five channels on a secondary basis with a maximum effective radiated power of 50 W PEP relative to a half wave dipole. Only upper sideband suppressed carrier voice transmissions may be used. The frequencies are 5330.5, 5346.5, 5366.5, 5371.5 and 5403.5 kHz. The occupied bandwidth is limited to 2.8 kHz centered on 5332, 5348, 5368, 5373, and 5405 kHz respectively.

40 Meters (7 MHz)
7000 — 7025 — 7125 — 7175 — 7300 kHz
7125 — N,T (200 W)
E† / A† / G†

† Phone and Image modes are permitted between 7075 and 7100 kHz for FCC licensed stations in ITU Regions 1 and 3 and by FCC licensed stations in ITU Region 2 West of 130 degrees West longitude or South of 20 degrees North latitude. See Sections 97.305(c) and 97.307(f)(11). Novice and Technician licensees outside ITU Region 2 may use CW only between 7025 and 7075 kHz. See Section 97.301(e). These exemptions do not apply to stations in the continental US.

30 Meters (10.1 MHz)
Avoid interference to fixed services outside the US.
10,100 — 10,150 kHz — E,A,G (200 Watts PEP)

20 Meters (14 MHz)
14,000 — 14,150 — 14,175 — 14,225 — 14,350 kHz
E / A / G
14,025 — E,A,G

17 Meters (18 MHz)
18,068 — 18,110 — 18,168 kHz — E,A,G

15 Meters (21 MHz)
21,000 — 21,200 — 21,225 — 21,275 — 21,450 kHz
E / A / G
21,025 — 21,200 — N,T (200 W)

12 Meters (24 MHz)
24,890 — 24,930 — 24,990 kHz — E,A,G

10 Meters (28 MHz)
28,000 — 28,300 — 28,500 — 29,700 kHz — E,A,G
28,000 — N,T (200 W)

6 Meters (50 MHz)
50.0 — 50.1 — 54.0 MHz — E,A,G,T

2 Meters (144 MHz)
144.0 — 144.1 — 148.0 MHz — E,A,G,T

1.25 Meters (222 MHz)
219.0 — 220.0 — 222.0 — 225.0 MHz
N (25 Watts) / E,A,G,T

*Geographical and power restrictions may apply to all bands above 420 MHz. See *The ARRL Operating Manual* for information about your area.

70 cm (420 MHz)*
420.0 — 450.0 MHz — E,A,G,T

33 cm (902 MHz)*
902.0 — 928.0 MHz — E,A,G

23 cm (1240 MHz)*
1240 — 1270 — 1295 — 1300 MHz — E,A,G,T
N (5 Watts)

All licensees except Novices are authorized all modes on the following frequencies:
- 2300-2310 MHz
- 2390-2450 MHz
- 3300-3500 MHz
- 5650-5925 MHz
- 10.0-10.5 GHz
- 24.0-24.25 GHz
- 47.0-47.2 GHz
- 76.0-81.0 GHz
- 122.25-123.0 GHz
- 134-141 GHz
- 241-250 GHz
- All above 275 GHz

ARRL — We're At Your Service

ARRL Headquarters: 860-594-0200 (Fax 860-594-0259)
email: hq@arrl.org

Publication Orders: www.arrl.org/catalog
Toll-Free 1-888-277-5289 (860-594-0355)
email: orders@arrl.org

Membership/Circulation Desk: www.arrl.org/join
Toll-Free 1-888-277-5289 (860-594-0338)
email: membership@arrl.org

Getting Started in Amateur Radio:
Toll-Free 1-800-326-3942 (860-594-0355)
email: newham@arrl.org

Exams: 860-594-0300 email: vec@arrl.org

Copyright © 2007, ARRL rev. 3/26/2007

Fig 2 — US amateur operating privileges.

Emergency communications is important enough that it warrants its own section, Subpart E. Amateur Radio volunteers may be found out in the field, or they may assist at Emergency Operations Centers, like this one activated in the aftermath of a hurricane.

Volunteer Examiner (VE) teams make it possible for you to take license exams. Whether you're going for your first license or an upgrade, you can find the nearest exam session on the ARRL website at www.arrl.org/arrlvec/examsearch.phtml.

call sign at least every 10 minutes during the contact and at the end of your communications.

SUBPART F — QUALIFYING EXAMINATION SYSTEMS

As discussed previously, the FCC delegates most of the Amateur Radio examination process to the amateur community itself, through the Volunteer Examiner (VE) program. The FCC maintains the standards for what classes of license are available in the US [§97.501] and what is necessary to pass the various examinations [§97.503]. The actual content, preparation and administration of exams are determined by Volunteer Examiner Coordinators (VECs).

The individual VECs are responsible for training and certifying qualified Amateur Radio operators to serve as Volunteer Examiners. They are also responsible for implementing the FCC guidelines for administration of exams and filing the results of those exams electronically with the FCC.

The individual VECs across the US work jointly through the National Conference of Volunteer Examiner Coordinators (NCVEC). The NCVEC is responsible, through its Question Pool Committee, to maintain a common question pool for each license class [§97.523].

SUMMARY

The wide variety of Amateur Radio bands, modes and activities available to you can be overwhelming at times. At every step you have to ensure that you are operating your station legally, safely and appropriately — no small task! While the regulations can seem daunting, the answers to most rules and regulations questions and issues can be determined from a good review of Part 97.

This chapter is not intended to provide an exhaustive discussion of every rule and regulation. Rather, it's intended to answer the most commonly asked questions. Each amateur is encouraged to have a copy of the current rules handy.

SOME FREQUENTLY ASKED QUESTIONS

Q. What is the difference between being a Primary user and Secondary user on a band?

A. The US Frequency Allocations are designed to provide efficient use of spectrum under the guidelines of international treaties and FCC/NTIA policies. In many portions of the spectrum, one radio service may be designated as Primary with one or more other services being given Secondary status. Where a band has shared access, the Primary service is afforded protection from harmful interference caused by an authorized Secondary user.

(47 C.F.R. 2.105 (c) (2) Stations of a secondary service:

(i) Shall not cause harmful interference to stations of primary services to which frequencies are already assigned or to which frequencies may be assigned at a later date;

(ii) Cannot claim protection from harmful interference from stations of a primary service to which frequencies are already assigned or may be assigned at a later date;)

On most HF bands where there is an Amateur Radio allocation, we are the designated Primary user. The HF exception is the 60-meter band, where we hold a Secondary status. On VHF, the Amateur Service is Primary on the 6 and 2 meter bands, as well as some portions of the 1.25 meter band.

Because of our Secondary status on most of the UHF and SHF bands, it is important for us to keep in mind that if our operation causes harmful interference to a Primary user, we have obligation to eliminate the interference. We also cannot claim protection from any interference received from users who have Primary status on a given band.

Q. What is the difference between a frequency allocation and a band plan?

A. A *frequency allocation* is the list of frequencies and modes authorized to a licensee. It comes directly from FCC allocations stated in parts 97.301 and 97.305.

Band plans are voluntary recommendations that describe where certain permitted activities are encouraged to take place on a band. Band plans promote more efficient use of our limited spectrum and enable operators to find others with similar interests on the band. For example, Part 97 rules allow SSTV anywhere on the 10-meter band between 28.300 and 28.700 MHz. However, to allow SSTV enthusiasts to more easily find SSTV operations, and to allow a more efficient use of the spectrum, the ARRL Band Plan recommends SSTV on 10 meters to take place at or near the frequency of 28.680 MHz. This doesn't mean SSTV operation would be illegal on another phone frequency on 10 meters, but band plans help encourage a more courteous and orderly environment for all operators.

Q. I am a General class licensee and I received an Official Observer notice for operating SSB (upper sideband) at 14.349 MHz. My license class permits me to operate up to 14.350. Why was I cited for operating out of the band? How close to the band edge can I legally operate?

A. There are several factors that come into play. First and foremost, you need to remember that your carrier and all of your sidebands must be confined within the amateur bands and subbands for your license class.

Amateurs commonly consider a fully suppressed single side band signal to be about 3 kHz wide. So as a general rule of thumb, you need to stay at least 3 kHz from the band edge if operating SSB. However, if your signal is wider than 3 kHz you will need to adjust your frequency to accommodate that extra bandwidth. Of course, if your equipment is fine-tuned enough and you have measured your signal to be less than 3 kHz, you can move a bit closer to the band edge. But remember, those stations you wish to work also have to adhere to the same limit and their signal may not be as trim as yours.

In your specific instance, you were operating upper sideband at 14.349 MHz according to your transceiver frequency display. Your display indicates your suppressed carrier frequency—it doesn't tell you that your signal is really extending more than 1 kHz above that frequency. That's why you received the notice for operating out of band.

Q. There was a big discussion on my nightly net about how often a station must ID. What's the answer?

A. Part 97.119 states a station "must transmit its assigned call sign on its transmitting channel at the end of each communication, and at least every ten minutes during a communication, for the purpose of clearly making the source of the transmissions from the station known to those receiving the transmissions."

To be practical, when checking into a net or establishing contact with a station, you are likely to identify by sending/announcing your call sign. But after establishing contact, you need only send your assigned station ID every 10 minutes if you are actively involved in the communications.

Let's say you are part of a roundtable net. You stated your call sign when you joined the net, but you've been simply listening ever since. Now it is your turn to speak. If 10 or more minutes have passed, you must identify again. If not, you can begin talking without the need to ID.

Q. Our club will be holding a special event using the club's call sign. The call sign belongs to an Amateur Extra class licensee. I hold a Technician license. Will I be able to operate on Extra class frequencies during the event?

A. The answer depends on the class of the control operator in charge at any given time. A club or special event 1 × 1 call sign *does not convey any operating privileges*. Also, the license class of the trustee is immaterial in these cases. It all comes down to the license held by the control operator at that moment.

If the control operator is a Technician, then the station may only be operated on the Technician class frequencies. To operate the radio on Extra class frequencies, there must be an Extra class licensee *present at the control point* of the radio.

Q. Our club holds a weekly swap net on the local repeater. Is that legal?

A. Part 97.113 (a) (3) says in part "Amateur operators may, however, notify other amateur operators of the availability for sale or trade of apparatus normally used in an amateur station, provided that such activity is not conducted on a regular basis."

So swap nets on the air are legal, but with a couple of considerations. A swap net is intended to function as a platform for hams to communicate that they have equipment available for sale. Note the wording of 97.113 (a) (3) where it states "notify of availability for sale or trade." This means it is legal to list (announce) equipment that is for sale, but nothing more. You should not negotiate the terms of sale over the air. Also, a person who operates a commercial business involved in selling ham gear should not use a swap net to advertise equipment available for sale, even if it is used equipment.

Q. I am an employee at the local hospital and am also a licensed amateur. In a disaster situation, as part of my job, the hospital wants to assign me to handle Amateur Radio communications. Isn't that a violation of Part 97?

A. In the 2006 "Omnibus" FCC Report and Order for Docket 04-140, The FCC clarified this issue. The R&O reads in part: *"These individuals are not receiving compensation for transmitting amateur service communications; rather, they are receiving compensation for services related to their disaster relief duties and in their capacities as emergency personnel."*

So in a disaster relief situation, the hospital employee is being compensated for his or her emergency responder duties, not as a paid Amateur Radio operator. Remember also that this applies to actual emergency scenarios, not day to day routine operations at the facility.

Q. I travel frequently and enjoy operating while mobile. How do I legally identify my station when I am operating away from home?

A. If you are traveling within the United States or US territories, you are only obligated to identify with your FCC call sign. Many stations ID with a special appended call sign such as

"W1AW/m" to indicate mobile operation, "W1AW/p" to indicate portable, or "W1AW/7" to indicate operating in a different call area. This is strictly optional; the FCC has not required this sort of indicator for many years. Many US licensees will sign with /KH6 if operating portable in Hawaii to let others know where they are located, but this is not required by the rules.

Under the rules, a foreign licensee operating in the US under one of the various reciprocal operating agreements identifies by appending their call sign with the letter W followed by the number of the US call district where they are located, such as "W1/IKØHBN."

Along the same lines, "maritime mobile" indicates you are operating on the high seas in international waters. You wouldn't identify as /MM while sailing your boat up the US Intercoastal Waterway or on an inland US lake, since it is still in US territory. You would ID as /MM when you are outside of the territorial waters of a county. If you are within the territorial waters of a specific country, you would need to have an operating permit with that country. Simply being on a US-flagged vessel doesn't apply if that ship is in another country's territorial waters.

Q. My friend, a non-amateur from Peru, is visiting the US and would like to talk to his brother back in Lima who is a licensed amateur. Is that legal third-party traffic? What if I want to talk to my non-amateur mom in Seattle while I am traveling in Germany on business?

A. An up-to-date list of countries with which we have third-party agreements can be found on the ARRLWeb at **www.arrl.org/FandES/field/regulations/io/3rdparty.html**.

It would be permissible for you to allow your Peruvian friend to participate as a third party and talk to his brother in Peru because we have a current third-party agreement with that country. However, you could not use Amateur Radio to have a conversation with your mom while in Germany because we do not have a third-party agreement with Germany.

Of course, in an emergency Part 97.115 allows US amateurs to conduct third-party traffic when transmitting emergency or disaster relief communications—regardless of the country in question.

Q. As a public service, the local TV station wants to set up an Amateur Radio weather spotting network that it would use to supplement their weather reporting during severe weather situations. Is this legal?

A. Simply put, no. According to Part 97.113, Amateur Radio is prohibited from being used to gather news for broadcast purposes. While the TV station can monitor the Amateur Radio net, it is not allowed to actively participate and use the amateur stations to gather information for their broadcasts. The exception to this would be if there was no other means of dissemination of the information to the public and the information is necessary for the immediate protection of life and property. Amateur Radio is also not permitted to be used as a routine broadcast radio service.

Subpart A—General Provisions

§97.1 Basis and purpose.
The rules and regulations in this Part are designed to provide an amateur radio service having a fundamental purpose as expressed in the following principles:
(a) Recognition and enhancement of the value of the amateur service to the public as a voluntary noncommercial communication service, particularly with respect to providing emergency communications.
(b) Continuation and extension of the amateur's proven ability to contribute to the advancement of the radio art.
(c) Encouragement and improvement of the amateur service through rules which provide for advancing skills in both the communications and technical phases of the art.
(d) Expansion of the existing reservoir within the amateur radio service of trained operators, technicians, and electronics experts.
(e) Continuation and extension of the amateur's unique ability to enhance international goodwill.

§97.3 Definitions.
(a) The definitions of terms used in Part 97 are:
    (1) Amateur operator. A person named in an amateur operator/primary license station grant on the ULS consolidated licensee database to be the control operator of an amateur station.
    (2) Amateur radio services. The amateur service, the amateur-satellite service and the radio amateur civil emergency service.
    (4) Amateur service. A radiocommunication service for the purpose of self-training, intercommunication and technical investigations carried out by amateurs, that is, duly authorized persons interested in radio technique solely with a personal aim and without pecuniary interest.
    (5) Amateur station. A station in an amateur radio service consisting of the apparatus necessary for carrying on radiocommunications.
    (6) Automatic control. The use of devices and procedures for control of a station when it is transmitting so that compliance with the FCC Rules is achieved without the control operator being present at a control point.
    (7) Auxiliary station. An amateur station, other than in a message forwarding system, that is transmitting communications point-to-point within a system of cooperating amateur stations.
    (8) Bandwidth. The width of a frequency band outside of which the mean power of the transmitted signal is attenuated at least 26 dB below the mean power of the transmitted signal within the band.
    (9) Beacon. An amateur station transmitting communications for the purposes of observation of propagation and reception or other related experimental activities.
    (10) Broadcasting. Transmissions intended for reception by the general public, either direct or relayed.
    (11) Call sign system. The method used to select a call sign for amateur station over-the-air identification purposes. The call sign systems are:
        (i) Sequential call sign system. The call sign is selected by the FCC from an alphabetized list corresponding to the geographic region of the licensee's mailing address and operator class. The call sign is shown on the license. The FCC will issue public announcements detailing the procedures of the sequential call sign system.
        (ii) Vanity call sign system. The call sign is selected by the FCC from a list of call signs requested by the licensee. The call sign is shown on the license. The FCC will issue public announcements detailing the procedures of the vanity call sign system.
        (iii) Special event call sign system. The call sign is selected by the station licensee from a list of call signs shown on a common data base coordinated, maintained and disseminated by the amateur station special event call sign data base coordinators. The call sign must have the single letter prefix K, N or W, followed by a single numeral 0 through 9, followed by a single letter A through W or Y or Z (for example K1A). The special event call sign is substituted for the call sign shown on the station license grant while the station is transmitting. The FCC will issue public announcements detailing the procedures of the special event call sign system.
    (12) CEPT radio amateur license. A license issued by a country belonging to the European Conference of Postal and Telecommunications Administrations (CEPT) that has adopted Recommendation T/R 61-01 (Nice 1985, Paris 1992, Nicosia 2003).
    (13) Control operator. An amateur operator designated by the licensee of a station to be responsible for the transmissions from that station to assure compliance with the FCC Rules.
    (14) Control point. The location at which the control operator function is performed.
    (15) CSCE. Certificate of successful completion of an examination.
    (16) Earth station. An amateur station located on, or within 50 km of the Earth's surface intended for communications with space stations or with other Earth stations by means of one or more other objects in space.
    (17) [Reserved]

(18) External RF Power Amplifier. A device capable of increasing power output when used in conjunction with, but not an integral part of, a transmitter.
(19) [Reserved]
(20) FAA. Federal Aviation Administration.
(21) FCC. Federal Communications Commission.
(22) Frequency coordinator. An entity, recognized in a local or regional area by amateur operators whose stations are eligible to be auxiliary or repeater stations, that recommends transmit/receive channels and associated operating and technical parameters for such stations in order to avoid or minimize potential interference.
(23) Harmful interference. Interference which endangers the functioning of a radionavigation service or of other safety services or seriously degrades, obstructs or repeatedly interrupts a radiocommunication service operating in accordance with the Radio Regulations.
(24) IARP. International Amateur Radio Permit. A document issued pursuant to the terms of the Inter-American Convention on an International Amateur Radio Permit by a country signatory to that Convention, other than the United States. Montrouis, Haiti.
(25) Indicator. Words, letters or numerals appended to and separated from the call sign during the station identification.
(26) Information bulletin. A message directed only to amateur operators consisting solely of subject matter of direct interest to the amateur service.
(27) International Morse code. A dot-dash code as defined in ITU-T Recommendation F.1 (March, 1998), Division B, I. Morse code.
(28) ITU. International Telecommunication Union.
(29) Line A. Begins at Aberdeen, WA, running by great circle arc to the intersection of 48° N, 120° W, thence along parallel 48° N, to the intersection of 95° W, thence by great circle arc through the southernmost point of Duluth, MN, thence by great circle arc to 45° N, 85° W, thence southward along meridian 85° W, to its intersection with parallel 41° N, thence along parallel 41° N, to its intersection with meridian 82° W, thence by great circle arc through the southernmost point of Bangor, ME, thence by great circle arc through the southernmost point of Searsport, ME, at which point it terminates.
(30) Local control. The use of a control operator who directly manipulates the operating adjustments in the station to achieve compliance with the FCC Rules.
(31) Message forwarding system. A group of amateur stations participating in a voluntary, cooperative, interactive arrangement where communications are sent from the control operator of an originating station to the control operator of one or more destination stations by one or more forwarding stations.
(32) National Radio Quiet Zone. The area in Maryland, Virginia and West Virginia bounded by 39° 15' N on the north, 78° 30' W on the east, 37° 30' N on the south and 80° 30' W on the west.
(33) Physician. For the purposes of this Part, a person who is licensed to practice in a place where the amateur service is regulated by the FCC, as either a Doctor of Medicine (MD) or a Doctor of Osteopathy (DO).
(34) Question pool. All current examination questions for a designated written examination element.
(35) Question set. A series of examination questions on a given examination selected from the question pool.
(36) Radio Regulations. The latest ITU *Radio Regulations* to which the United States is a party.
(37) RACES (radio amateur civil emergency service). A radio service using amateur stations for civil defense communications during periods of local, regional or national civil emergencies.
(38) Remote control. The use of a control operator who indirectly manipulates the operating adjustments in the station through a control link to achieve compliance with the FCC Rules.
(39) Repeater. An amateur station that simultaneously retransmits the transmission of another amateur station on a different channel or channels.
(40) Space station. An amateur station located more than 50 km above the Earth's surface.
(41) Space telemetry. A one-way transmission from a space station of measurements made from the measuring instruments in a spacecraft, including those relating to the functioning of the spacecraft.
(42) Spurious emission. An emission, on frequencies outside the necessary bandwidth of a transmission, the level of which may be reduced without affecting the information being transmitted.
(43) Telecommand. A one-way transmission to initiate, modify, or terminate functions of a device at a distance.
(44) Telecommand station. An amateur station that transmits communications to initiate, modify, or terminate functions of a space station.
(45) Telemetry. A one-way transmission of measurements at a distance from the measuring instrument.
(46) Third-party communications. A message from the control operator (first party) of an amateur station to another amateur station control operator (second party) on behalf of another person (third party).
(47) ULS (Universal Licensing System). The consolidated database, application filing system and processing system for all Wireless Telecommunications Services.
(48) VE. Volunteer examiner.

(49) *VEC*. Volunteer-examiner coordinator.

(b) The definitions of technical symbols used in this Part are:

(1) *EHF* (extremely high frequency). The frequency range 30-300 GHz.

(2) *HF* (high frequency). The frequency range 3-30 MHz.

(3) *Hz*. Hertz.

(4) *m*. Meters.

(5) *MF* (medium frequency). The frequency range 300-3000 kHz.

(6) *PEP* (peak envelope power). The average power supplied to the antenna transmission line by a transmitter during one RF cycle at the crest of the modulation envelope taken under normal operating conditions.

(7) *RF*. Radio frequency.

(8) *SHF* (super-high frequency). The frequency range 3-30 GHz.

(9) *UHF* (ultra-high frequency). The frequency range 300-3000 MHz.

(10) *VHF* (very-high frequency). The frequency range 30-300 MHz.

(11) *W*. Watts.

(c) The following terms are used in this Part to indicate emission types. Refer to Â§2.201 of the FCC Rules, *Emission, modulation and transmission characteristics*, for information on emission type designators.

(1) *CW*. International Morse code telegraphy emissions having designators with A, C, H, J or R as the first symbol; 1 as the second symbol; A or B as the third symbol; and emissions J2A and J2B.

(2) *Data*. Telemetry, telecommand and computer communications emissions having (i) designators with A, C, D, F, G, H, J or R as the first symbol, 1 as the second symbol, and D as the third symbol; (ii) emission J2D; and (iii) emissions A1C, F1C, F2C, J2C, and J3C having an occupied bandwidth of 500 Hz or less when transmitted on an amateur service frequency below 30 MHz. Only a digital code of a type specifically authorized in this part may be transmitted.

(3) *Image*. Facsimile and television emissions having designators with A, C, D, F, G, H, J or R as the first symbol; 1, 2 or 3 as the second symbol; C or F as the third symbol; and emissions having B as the first symbol; 7, 8 or 9 as the second symbol; W as the third symbol.

(4) *MCW*. Tone-modulated international Morse code telegraphy emissions having designators with A, C, D, F, G, H or R as the first symbol; 2 as the second symbol; A or B as the third symbol.

(5) *Phone*. Speech and other sound emissions having designators with A, C, D, F, G, H, J or R as the first symbol; 1, 2 or 3 as the second symbol; E as the third symbol. Also speech emissions having B as the first symbol; 7, 8 or 9 as the second symbol; E as the third symbol. MCW for the purpose of performing the station identification procedure, or for providing telegraphy practice interspersed with speech. Incidental tones for the purpose of selective calling or alerting or to control the level of a demodulated signal may also be considered phone.

(6) *Pulse*. Emissions having designators with K, L, M, P, Q, V or W as the first symbol; 0, 1, 2, 3, 7, 8, 9 or X as the second symbol; A, B, C, D, E, F, N, W or X as the third symbol.

(7) *RTTY*. Narrow-band direct-printing telegraphy emissions having designators with A, C, D, F, G, H, J or R as the first symbol; 1 as the second symbol; B as the third symbol; and emission J2B. Only a digital code of a type specifically authorized in this part may be transmitted.

(8) *SS*. Spread-spectrum emissions using bandwidth-expansion modulation emissions having designators with A, C, D, F, G, H, J or R as the first symbol; X as the second symbol; X as the third symbol.

(9) *Test*. Emissions containing no information having the designators with N as the third symbol. Test does not include pulse emissions with no information or modulation unless pulse emissions are also authorized in the frequency band.

§97.5 Station license grant required.

(a) The station apparatus must be under the physical control of a person named in an amateur station license grant on the ULS consolidated license database or a person authorized for alien reciprocal operation by §97.107 of this part, before the station may transmit on any amateur service frequency from any place that is:

(1) Within 50 km of the Earth's surface and at a place where the amateur service is regulated by the FCC;

(2) Within 50 km of the Earth's surface and aboard any vessel or craft that is documented or registered in the United States; or

(3) More than 50 km above the Earth's surface aboard any craft that is documented or registered in the United States.

(b) The types of station license grants are:

(1) An operator/primary station license grant. One, but only one, operator/primary station license grant may be held by any one person. The primary station license is granted together with the amateur operator license. Except for a representative of a foreign government, any person who qualifies by examination is eligible to apply for an operator/primary station license grant.

(2) A club station license grant. A club station license grant may be held only by the person who is the license trustee designated by an officer of the club. The trustee must be a person who holds an Amateur Extra, Advanced, General, or Technician operator license grant. The club must be composed of at least four persons and must have a name, a document of organization, management, and a primary purpose devoted to amateur service activities consistent with this part.

(3) A military recreation station license grant. A military recreation station license grant may be held only by the person who is the license custodian designated by the official in charge of the United States military recreational premises where the station is situated. The person must not be a representative of a foreign government. The person need not hold an amateur operator license grant.

(4) A RACES station license grant. A RACES station license grant may be held only by the person who is the license custodian designated by the official responsible for the governmental agency served by that civil defense organization. The custodian must be the civil defense official responsible for coordination of all civil defense activities in the area concerned. The custodian must not be a representative of a foreign government. The custodian need not hold an amateur operator license grant.

(c) The person named in the station license grant or who is authorized for alien reciprocal operation by §97.107 of this Part may use, in accordance with the applicable rules of this Part, the transmitting apparatus under the physical control of the person at places where the amateur service is regulated by the FCC.

(d) A CEPT radio-amateur license is issued to the person by the country of which the person is a citizen. The person must not:

(1) Be a resident alien or citizen of the United States, regardless of any other citizenship also held;

(2) Hold an FCC-issued amateur operator license nor reciprocal permit for alien amateur licensee;

(3) Be a prior amateur service licensee whose FCC-issued license was revoked, suspended for less than the balance of the license term and the suspension is still in effect, suspended for the balance of the license term and relicensing has not taken place, or surrendered for cancellation following notice of revocation, suspension or monetary forfeiture proceedings; or

(4) Be the subject of a cease and desist order that relates to amateur service operation and which is still in effect.

(e) An IARP is issued to the person by the country of which the person is a citizen. The person must not:

(1) Be a resident alien or citizen of the United States, regardless of any other citizenship also held;

(2) Hold an FCC-issued amateur operator license nor reciprocal permit for alien amateur licensee;

(3) Be a prior amateur service licensee whose FCC-issued license was revoked, suspended for less than the balance of the license term and the suspension is still in effect, suspended for the balance of the license term and relicensing has not taken place, or surrendered for cancellation following notice of revocation, suspension or monetary forfeiture proceedings; or

(4) Be the subject of a cease and desist order that relates to amateur service operation and which is still in effect.

§97.7 Control operation required.

When transmitting, each amateur station must have a control operator. The control operator must be a person:
(a) For whom an amateur operator/primary station license grant appears on the ULS consolidated licensee database, or
(b) Who is authorized for alien reciprocal operation by §97.107 of this part

§97.9 Operator license grant.

(a) The classes of amateur operator license grants are: Novice, Technician, Technician Plus (until such licenses expire, a Technical Class license granted before February 14, 1991, is considered a Technician Plus Class license), General, Advanced, and Amateur Extra. The person named in the operator license grant is authorized to be the control operator of an amateur station with the privileges authorized to the operator class specified on the license grant.

(b) The person named in an operator license grant of Novice, Technician, Technician Plus, General or Advanced Class, who has properly submitted to the administering VEs, a FCC Form 605 document requesting examination for an operator license grant of a higher class, and who holds a CSCE indicating that the person has completed the necessary examinations within the previous 365 days, is authorized to exercise the rights and privileges of the higher operator class until a final disposition of the application or until 365 days following the passing of the examination, whichever comes first.

§97.11 Stations aboard ships or aircraft.

(a) The installation and operation of an amateur station on a ship or aircraft must be approved by the master of the ship or pilot in command of the aircraft.

(b) The station must be separate from and independent of all other radio apparatus installed on the ship or aircraft, except a common antenna may be shared with a voluntary ship radio installation. The station's transmissions must not cause interference to any other apparatus installed on the ship or aircraft.

(c) The station must not constitute a hazard to the safety of life or property. For a station aboard an aircraft, the apparatus shall not be operated while the aircraft is operating under Instrument Flight Rules, as defined by the FAA, unless the station has been found to comply with all applicable FAA Rules.

§97.13 Restrictions on station location.

(a) Before placing an amateur station on land of environmental importance or that is significant in American history, architecture or culture, the licensee may be required to take certain actions prescribed by §§ 1.1305-1.1319 of this chapter.

(b) A station within 1600 m (1 mile) of an FCC monitoring facility must protect that facility from harmful interference. Failure to do so could result in imposition of operating restrictions upon the amateur station by a District Director pursuant to §97.121 of this Part. Geographical coordinates of the facilities that require protection are listed in §0.121(c) of this chapter.

(c) Before causing or allowing an amateur station to transmit from any place where the operation of the station could cause human exposure to RF electromagnetic field levels in excess of those allowed under § 1.1310 of this chapter, the licensee is required to take certain actions.

(1) The licensee must perform the routine RF environmental evaluation prescribed by § 1.1307(b) of this chapter, if the power of the licensee's station exceeds the limits given in the following table:

Wavelength Band	Evaluation Required if Power* (watts) Exceeds:
MF	
160m	500
HF	
80m	500
75m	500
40m	500
30m	425
20m	225
17m	125
15m	100
12m	75
10m	50
VHF (all bands)	50
UHF	
70cm	70
33cm	150
23cm	200
13cm	250
SHF (all bands)	250
EHF (all bands)	250

Repeater stations (all bands) *non-building-mounted antennas:* height above ground level to lowest point of antenna < 10 m *and* power > 500 W ERP *building-mounted antennas:* power > 500 W ERP

* Power = PEP input to antenna except, for repeater stations only, power exclusion is based on ERP (effective radiated power).

(2) If the routine environmental evaluation indicates that the RF electromagnetic fields could exceed the limits contained in § 1.1310 of this chapter in accessible areas, the licensee must take action to prevent human exposure to such RF electromagnetic fields. Further information on evaluating compliance with these limits can be found in the FCC's OET Bulletin Number 65, "Evaluating Compliance with FCC-Specified Guidelines for Human Exposure to Radio Frequency Electromagnetic Fields."

§97.15 Station antenna structures.

(a) Owners of certain antenna structures more than 60.96 meters (200 feet) above ground level at the site or located near or at a public use airport must notify the Federal Aviation Administration and register with the Commission as required by Part 17 of this chapter.

(b) Except as otherwise provided herein, a station antenna structure may be erected at heights and dimensions sufficient to accommodate amateur service communications. [State and local regulation of a station antenna structure must not preclude amateur service communications. Rather, it must reasonably accommodate such communications and must constitute the minimum practicable regulation to accomplish the state or local authority's legitimate purpose. See PRB-1, 101 FCC 2d 952 (1985) for details.]

§97.17 Application for new license grant.

(a) Any qualified person is eligible to apply for a new operator/primary station, club station or military recreation station license grant. No new license grant will be issued for a Novice, Technician Plus, or Advanced Class operator/primary station or a RACES station.

(b) Each application for a new amateur service license grant must be *filed* with the FCC as follows:

(1) Each candidate for an amateur radio operator license which requires the applicant to pass one or more examination elements must present the administering VEs with all information required by the rules prior to the examination. The VEs may collect all necessary information in any manner of their choosing, including creating their own forms.

(2) For a new club or military recreation station license grant, each applicant must present all information required by the rules to an amateur radio organization having tax-exempt status under section 501(c)(3) of the Internal Revenue

Code of 1986 that provides voluntary, uncompensated and unreimbursed services in providing club and military recreation station call signs ("Club Station Call Sign Administrator") who must submit the information to the FCC in an electronic batch file. The Club Station Call Sign Administrator may collect the information required by these rules in any manner of their choosing, including creating their own forms. The Club Station Call Sign Administrator must retain the applicants information for at least 15 months and make it available to the FCC upon request. The FCC will issue public announcements listing the qualified organizations that have completed a pilot autogrant batch filing project and are authorized to serve as a Club Station Call Sign Administrator.

(c) No person shall obtain or attempt to obtain, or assist another person to obtain or attempt to obtain, an amateur service license grant by fraudulent means.

(d) One unique call sign will be shown on the license grant of each new primary, club and military recreation station. The call sign will be selected by the sequential call sign system.

§97.19 Application for a vanity call sign.

(a) The person named in an operator/primary station license grant or in a club station license grant is eligible to make application for modification of the license grant, or the renewal thereof, to show a call sign selected by the vanity call sign system. RACES and military recreation stations are not eligible for a vanity call sign.

(b) Each application for a modification of an operator/primary or club station license grant, or the renewal thereof, to show a call sign selected by the vanity call sign system must be filed in accordance with §1.913 of this chapter.

(c) Unassigned call signs are available to the vanity call sign system with the following exceptions:

(1) A call sign shown on an expired license grant is not available to the vanity call sign system for 2 years following the expiration of the license.

(2) A call sign shown on a surrendered, revoked, set aside, canceled, or voided license grant is not available to the vanity call sign system for 2 years following the date such action is taken.

(3) Except for an applicant who is the spouse, child, grandchild, stepchild, parent, grandparent, step-parent, brother, sister, stepbrother, stepsister, aunt, uncle, niece, nephew, or in-law, and except for an applicant who is a club station license trustee acting with a written statement of consent signed by either the licensee ante mortem but who is now deceased or by at least one relative, as listed above, of the person now deceased, the call sign shown on the license of the person now deceased is not available to the vanity call sign system for 2 years following the person's death, or for 2 years following the expiration of the license grant, whichever is sooner.

(d) The vanity call sign requested by an applicant must be selected from the group of call signs corresponding to the same or lower class of operator license held by the applicant as designated in the sequential call sign system.

(1) The applicant must request that the call sign shown on the license grant be vacated and provide a list of up to 25 call signs in order of preference. In the event that the Commission receives more than one application requesting a vanity call sign from an applicant on the same receipt day, the Commission will process only the first such application entered into the Universal Licensing System. Subsequent vanity call sign applications from that applicant with the same receipt date will not be accepted.

(2) The first assignable call sign from the applicant's list will be shown on the license grant. When none of those call signs are assignable, the call sign vacated by the applicant will be shown on the license grant.

(3) Vanity call signs will be selected from those call signs assignable at the time the application is processed by the FCC.

(4) A call sign designated under the sequential call sign system for Alaska, Hawaii, Caribbean Insular Areas, and Pacific Insular areas will be assigned only to a primary or club station whose licensee's mailing address is in the corresponding state, commonwealth, or island. This limitation does not apply to an applicant for the call sign as the spouse, child, grandchild, stepchild, parent, grandparent, stepparent, brother, sister, stepbrother, stepsister, aunt, uncle, niece, nephew, or in-law, of the former holder now deceased.

§97.21 Application for a modified or renewed license grant.

(a) A person holding a valid amateur station license grant:

(1) Must apply to the FCC for a modification of the license grant as necessary to show the correct mailing address, licensee name, club name, license trustee name or license custodian name in accordance with §1.913 of this chapter. For a club, military recreation or RACES station license grant, it must be presented in document form to a Club Station Call Sign Administrator who must submit the information thereon to the FCC in an electronic batch file. The Club Station Call Sign Administrator must retain the collected information for at least 15 months and make it available to the FCC upon request.

(2) May apply to the FCC for a modification of the operator/primary station license grant to show a higher operator class. Applicants must present the administering VEs with all information required by the rules prior to the examination. The VEs may collect all necessary information in any manner of their choosing, including creating their own forms.

(3) May apply to the FCC for renewal of the license grant for another term in accordance with §1.913 of this chapter. Application for renewal of a Technician Plus class operator/primary station license will be processed as an application for renewal of a Technician operator/primary station license.

(i) For a station license grant showing a call sign obtained through the vanity call sign system, the application must be filed in accordance with §97.19 of this Part in order to have the vanity call sign reassigned to the station.

(ii) For a primary station license grant showing a call sign obtained through the sequential call sign system, and for a primary station license grant showing a call sign obtained through the vanity call sign system but whose grantee does not want to have the vanity call sign reassigned to the station, the application must be filed with the FCC in accordance with §1.913 of this chapter. When the application has been received by the FCC on or before the license expiration date, the license operating authority is continued until the final disposition of the application.

(iii) For a club station or military recreation station license grant showing a call sign obtained through the sequential call sign system, and for a club or military recreation station license grant showing a call sign obtained through the vanity call sign system but whose grantee does not want to have the vanity call sign reassigned to the station, the application must be presented in document form to a Club Station Call Sign Administrator who must submit the information thereon to the FCC in an electronic batch file. The Club Station Call Sign Administrator must retain the collected information for at least 15 months and make it available to the FCC upon request. RACES station license grants will not be renewed.

(b) A person whose amateur station license grant has expired may apply to the FCC for renewal of the license grant for another term during a 2 year filing grace period. The application must be received at the address specified above prior to the end of the grace period. Unless and until the license grant is renewed, no privileges in this Part are conferred.

(c) A call sign obtained under the sequential or vanity call sign system will be reassigned to the station upon renewal or modification of a station license.

§97.23 Mailing address.

Each license grant must show the grantee's correct name and mailing address. The mailing address must be in an area where the amateur service is regulated by the FCC and where the grantee can receive mail delivery by the United States Postal Service. Revocation of the station license or suspension of the operator license may result when correspondence from the FCC is returned as undeliverable because the grantee failed to provide the correct mailing address.

§97.25 License term.

An amateur service license is normally granted for a 10-year term.

§97.27 FCC modification of station license grant.

(a) The FCC may modify a station license grant, either for a limited time or for the duration of the term thereof, if it determines:

    (1) That such action will promote the public interest, convenience, and necessity; or

    (2) That such action will promote fuller compliance with the provisions of the Communications Act of 1934, as amended, or of any treaty ratified by the United States.

(b) When the FCC makes such a determination, it will issue an order of modification. The order will not become final until the licensee is notified in writing of the proposed action and the grounds and reasons therefor. The licensee will be given reasonable opportunity of no less than 30 days to protest the modification; except that, where safety of life or property is involved, a shorter period of notice may be provided. Any protest by a licensee of an FCC order of modification will be handled in accordance with the provisions of 47 U.S.C. §316.

§97.29 Replacement license grant document.

Each grantee whose amateur station license grant document is lost, mutilated or destroyed may apply to the FCC for a replacement in accordance with §1.913 of this chapter.

Subpart B—Station Operation Standards

§97.101 General standards.
(a) In all respects not specifically covered by FCC Rules each amateur station must be operated in accordance with good engineering and good amateur practice.
(b) Each station licensee and each control operator must cooperate in selecting transmitting channels and in making the most effective use of the amateur service frequencies. No frequency will be assigned for the exclusive use of any station.
(c) At all times and on all frequencies, each control operator must give priority to stations providing emergency communications, except to stations transmitting communications for training drills and tests in RACES.
(d) No amateur operator shall willfully or maliciously interfere with or cause interference to any radio communication or signal.

§97.103 Station licensee responsibilities.
(a) The station licensee is responsible for the proper operation of the station in accordance with the FCC Rules. When the control operator is a different amateur operator than the station licensee, both persons are equally responsible for proper operation of the station.
(b) The station licensee must designate the station control operator. The FCC will presume that the station licensee is also the control operator, unless documentation to the contrary is in the station records.
(c) The station licensee must make the station and the station records available for inspection upon request by an FCC representative. When deemed necessary by a District Director to assure compliance with the FCC Rules, the station licensee must maintain a record of station operations containing such items of information as the District Director may require in accord with § 0.314(x) of the FCC Rules.

§97.105 Control operator duties.
(a) The control operator must ensure the immediate proper operation of the station, regardless of the type of control.
(b) A station may only be operated in the manner and to the extent permitted by the privileges authorized for the class of operator license held by the control operator.

§97.107 Reciprocal operating authority.
A non-citizen of the United States ("alien") holding an amateur service authorization granted by the alien's government is authorized to be the control operator of an amateur station located at places where the amateur service is regulated by the FCC, provided there is in effect a multilateral or bilateral reciprocal operating arrangement, to which the United States and the alien's government are parties, for amateur service operation on a reciprocal basis. The FCC will issue public announcements listing the countries with which the United States has such an arrangement. No citizen of the United States or person holding an FCC amateur operator/primary station license grant is eligible for the reciprocal operating authority granted by this section. The privileges granted to a control operator under this authorization are:
(a) For an amateur service license granted by the Government of Canada:
    (1) The terms of the *Convention Between the United States and Canada (TIAS No. 2508) Relating to the Operation by Citizens of Either Country of Certain Radio Equipment or Stations in the Other Country*;
    (2) The operating terms and conditions of the amateur service license issued by the Government of Canada; and
    (3) The applicable rules of this part, but not to exceed the control operator privileges of an FCC-granted Amateur Extra Class operator license.
(b) For an amateur service license granted by any country, other than Canada, with which the United States has a multilateral or bilateral agreement:
    (1) The terms of the agreement between the alien's government and the United States;
    (2) The operating terms and conditions of the amateur service license granted by the alien's government;
    (3) The applicable rules of this part, but not to exceed the control operator privileges of an FCC-granted Amateur Extra Class operator license; and
(c) At any time the FCC may, in its discretion, modify, suspend or cancel the reciprocal operating authority granted to any person by this section.

§97.109 Station control.
(a) Each amateur station must have at least one control point.
(b) When a station is being locally controlled, the control operator must be at the control point. Any station may be locally controlled.
(c) When a station is being remotely controlled, the control operator must be at the control point. Any station may be remotely controlled.
(d) When a station is being automatically controlled, the control operator need not be at the control point. Only stations specifically designated elsewhere in this Part may be automatically controlled. Automatic control must cease upon notification by

a District Director that the station is transmitting improperly or causing harmful interference to other stations. Automatic control must not be resumed without prior approval of the District Director.

(e) No station may be automatically controlled while transmitting third party communications, except a station transmitting a RTTY or data emission. All messages that are retransmitted must originate at a station that is being locally or remotely controlled.

§97.111 Authorized transmissions.

(a) An amateur station may transmit the following types of two-way communications:

(1) Transmissions necessary to exchange messages with other stations in the amateur service, except those in any country whose administration has notified the ITU that it objects to such communications. The FCC will issue public notices of current arrangements for international communications;

(2) Transmissions necessary to meet essential communication needs and to facilitate relief actions.

(3) Transmissions necessary to exchange messages with a station in another FCC-regulated service while providing emergency communications;

(4) Transmissions necessary to exchange messages with a United States government station, necessary to providing communications in RACES; and

(5) Transmissions necessary to exchange messages with a station in a service not regulated by the FCC, but authorized by the FCC to communicate with amateur stations. An amateur station may exchange messages with a participating United States military station during an Armed Forces Day Communications Test.

(b) In addition to one-way transmissions specifically authorized elsewhere in this Part, an amateur station may transmit the following types of one-way communications:

(1) Brief transmissions necessary to make adjustments to the station;

(2) Brief transmissions necessary to establishing two-way communications with other stations;

(3) Telecommand;

(4) Transmissions necessary to providing emergency communications;

(5) Transmissions necessary to assisting persons learning, or improving proficiency in, the international Morse code;

(6) Transmissions necessary to disseminate information bulletins;

(7) Transmissions of telemetry.

§97.113 Prohibited transmissions.

(a) No amateur station shall transmit:

(1) Communications specifically prohibited elsewhere in this Part;

(2) Communications for hire or for material compensation, direct or indirect, paid or promised, except as otherwise provided in these rules;

(3) Communications in which the station licensee or control operator has a pecuniary interest, including communications on behalf of an employer. Amateur operators may, however, notify other amateur operators of the availability for sale or trade of apparatus normally used in an amateur station, provided that such activity is not conducted on a regular basis;

(4) Music using a phone emission except as specifically provided elsewhere in this section; communications intended to facilitate a criminal act; messages encoded for the purpose of obscuring their meaning, except as otherwise provided herein; obscene or indecent words or language; or false or deceptive messages, signals or identification;

(5) Communications, on a regular basis, which could reasonably be furnished alternatively through other radio services.

(b) An amateur station shall not engage in any form of broadcasting, nor may an amateur station transmit one-way communications except as specifically provided in these rules; nor shall an amateur station engage in any activity related to program production or news gathering for broadcasting purposes, except that communications directly related to the immediate safety of human life or the protection of property may be provided by amateur stations to broadcasters for dissemination to the public where no other means of communication is reasonably available before or at the time of the event.

(c) A control operator may accept compensation as an incident of a teaching position during periods of time when an amateur station is used by that teacher as a part of classroom instruction at an educational institution.

(d) The control operator of a club station may accept compensation for the periods of time when the station is transmitting telegraphy practice or information bulletins, provided that the station transmits such telegraphy practice and bulletins for at least 40 hours per week; schedules operations on at least six amateur service MF and HF bands using reasonable measures to maximize coverage; where the schedule of normal operating times and frequencies is published at least 30 days in advance of the actual transmissions; and where the control operator does not accept any direct or indirect compensation for any other service as a control operator.

(e) No station shall retransmit programs or signals emanating from any type of radio station other than an amateur station, except propagation and weather forecast information intended for use by the general public and originated from United States Government stations, and communications, including incidental music, originating on United States Government frequencies between a manned spacecraft and its associated Earth stations. Prior approval for manned spacecraft communications

retransmissions must be obtained from the National Aeronautics and Space Administration. Such retransmissions must be for the exclusive use of amateur radio operators. Propagation, weather forecasts, and manned spacecraft communications retransmissions may not be conducted on a regular basis, but only occasionally, as an incident of normal amateur radio communications.
(f) No amateur station, except an auxiliary, repeater or space station, may automatically retransmit the radio signals of other amateur stations.

§97.115 Third party communications.
(a) An amateur station may transmit messages for a third party to:
    (1) Any station within the jurisdiction of the United States.
    (2) Any station within the jurisdiction of any foreign government when transmitting emergency or disaster relief communications and any station within the jurisdiction of any foreign government whose administration has made arrangements with the United States to allow amateur stations to be used for transmitting international communications on behalf of third parties. No station shall transmit messages for a third party to any station within the jurisdiction of any foreign government whose administration has not made such an arrangement. This prohibition does not apply to a message for any third party who is eligible to be a control operator of the station.
(b) The third party may participate in stating the message where:
    (1) The control operator is present at the control point and is continuously monitoring and supervising the third party's participation; and
    (2) The third party is not a prior amateur service licensee whose license was revoked or not renewed after hearing and re-licensing has not taken place; suspended for less than the balance of the license term and the suspension is still in effect; suspended for the balance of the license term and re-licensing has not taken place; or surrendered for cancellation following notice of revocation, suspension or monetary forfeiture proceedings. The third party may not be the subject of a cease and desist order which relates to amateur service operation and which is still in effect.
(c) No station may transmit third party communications while being automatically controlled except a station transmitting a RTTY or data emission.

§97.117 International communications.
Transmissions to a different country, where permitted, shall be shall be limited to communications incidental to the purposes of the amateur service and to remarks of a personal character.

§97.119 Station identification.
(a) Each amateur station, except a space station or telecommand station, must transmit its assigned call sign on its transmitting channel at the end of each communication, and at least every ten minutes during a communication, for the purpose of clearly making the source of the transmissions from the station known to those receiving the transmissions. No station may transmit unidentified communications or signals, or transmit as the station call sign, any call sign not authorized to the station.
(b) The call sign must be transmitted with an emission authorized for the transmitting channel in one of the following ways:
    (1) By a CW emission. When keyed by an automatic device used only for identification, the speed must not exceed 20 words per minute;
    (2) By a phone emission in the English language. Use of a phonetic alphabet as an aid for correct station identification is encouraged;
    (3) By a RTTY emission using a specified digital code when all or part of the communications are transmitted by a RTTY or data emission;
    (4) By an image emission conforming to the applicable transmission standards, either color or monochrome, of §73.682(a) of the FCC Rules when all or part of the communications are transmitted in the same image emission.
(c) One or more indicators may be included with the call sign. Each indicator must be separated from the call sign by the slant mark (/) or by any suitable word that denotes the slant mark. If an indicator is self-assigned, it must be included before, after, or both before and after, the call sign. No self-assigned indicator may conflict with any other indicator specified by the FCC Rules or with any prefix assigned to another country.
(d) When transmitting in conjunction with an event of special significance, a station may substitute for its assigned call sign a special event call sign as shown for that station for that period of time on the common data base coordinated, maintained and disseminated by the special event call sign data base coordinators. Additionally, the station must transmit its assigned call sign at least once per hour during such transmissions.
(e) When the operator license class held by the control operator exceeds that of the station licensee, an indicator consisting of the call sign assigned to the control operator's station must be included after the call sign.
(f) When the control operator who is exercising the rights and privileges authorized by §97.9(b) of this Part, an indicator must be included after the call sign as follows:
    (1) For a control operator who has requested a license modification from Novice to Technician Class: KT;

(2) For a control operator who has requested a license modification from Novice, Technician or Technician Plus Class to General Class: AG;

(3) For a control operator who has requested a license modification from Novice, Technician, Technician Plus, General, or Advanced Class operator to Amateur Extra Class: AE.

(g) When the station is transmitting under the authority of §97.107 of this part, an indicator consisting of the appropriate letter-numeral designating the station location must be included before the call sign that was issued to the station by the country granting the license. For an amateur service license granted by the Government of Canada, however, the indicator must be included after the call sign. At least once during each intercommunication, the identification announcement must include the geographical location as nearly as possible by city and state, commonwealth or possession.

§97.121 Restricted operation.

(a) If the operation of an amateur station causes general interference to the reception of transmissions from stations operating in the domestic broadcast service when receivers of good engineering design, including adequate selectivity characteristics, are used to receive such transmissions, and this fact is made known to the amateur station licensee, the amateur station shall not be operated during the hours from 8 p.m. to 10:30 p.m., local time, and on Sunday for the additional period from 10:30 a.m. until 1 p.m., local time, upon the frequency or frequencies used when the interference is created.

(b) In general, such steps as may be necessary to minimize interference to stations operating in other services may be required after investigation by the FCC.

Subpart C—Special Operations

§97.201 Auxiliary station.
(a) Any amateur station licensed to a holder of a Technician, Technician Plus, General, Advanced or Amateur Extra Class operator license may be an auxiliary station. A holder of a Technician, Technician Plus, General, Advanced or Amateur Extra Class operator license may be the control operator of an auxiliary station, subject to the privileges of the class of operator license held.
(b) An auxiliary station may transmit only on the 2 m and shorter wavelength bands, except the 144.0-144.5 MHz, 145.8-146.0 MHz, 219-220 MHz, 222.00-222.15 MHz, 431-433 MHz, and 435-438 MHz segments.
(c) Where an auxiliary station causes harmful interference to another auxiliary station, the licensees are equally and fully responsible for resolving the interference unless one station's operation is recommended by a frequency coordinator and the other station's is not. In that case, the licensee of the non-coordinated auxiliary station has primary responsibility to resolve the interference.
(d) An auxiliary station may be automatically controlled.
(e) An auxiliary station may transmit one-way communications.

§97.203 Beacon station.
(a) Any amateur station licensed to a holder of a Technician, Technician Plus, General, Advanced or Amateur Extra Class operator license may be a beacon. A holder of a Technician, Technician Plus, General, Advanced or Amateur Extra Class operator license may be the control operator of a beacon, subject to the privileges of the class of operator license held.
(b) A beacon must not concurrently transmit on more than 1 channel in the same amateur service frequency band, from the same station location.
(c) The transmitter power of a beacon must not exceed 100 W.
(d) A beacon may be automatically controlled while it is transmitting on the 28.20-28.30 MHz, 50.06-50.08 MHz, 144.275-144.300 MHz, 222.05-222.06 MHz, or 432.300-432.400 MHz segments, or on the 33 cm and shorter wavelength bands.
(e) Before establishing an automatically controlled beacon in the National Radio Quiet Zone or before changing the transmitting frequency, transmitter power, antenna height or directivity, the station licensee must give written notification thereof to the Interference Office, National Radio Astronomy Observatory, P.O. Box 2, Green Bank, WV 24944.
  (1) The notification must include the geographical coordinates of the antenna, antenna ground elevation above mean sea level (AMSL), antenna center of radiation above ground level (AGL), antenna directivity, proposed frequency, type of emission, and transmitter power.
  (2) If an objection to the proposed operation is received by the FCC from the National Radio Astronomy Observatory at Green Bank, Pocahontas County, WV, for itself or on behalf of the Naval Research Laboratory at Sugar Grove, Pendleton County, WV, within 20 days from the date of notification, the FCC will consider all aspects of the problem and take whatever action is deemed appropriate.
(f) A beacon must cease transmissions upon notification by a District Director that the station is operating improperly or causing undue interference to other operations. The beacon may not resume transmitting without prior approval of the District Director.
(g) A beacon may transmit one-way communications.

§97.205 Repeater station.
(a) Any amateur station licensed to a holder of a Technician, General, Advanced or Amateur Extra Class operator license may be a repeater. A holder of a Technician, General, Advanced or Amateur Extra Class operator license may be the control operator of a repeater, subject to the privileges of the class of operator license held.
(b) A repeater may receive and retransmit only on the 10 m and shorter wavelength frequency bands except the 28.0-29.5 MHz, 50.0-51.0 MHz, 144.0-144.5 MHz, 145.5-146.0 MHz, 222.00-222.15 MHz, 431.0-433.0 MHz and 435.0-438.0 MHz segments.
(c) Where the transmissions of a repeater cause harmful interference to another repeater, the two station licensees are equally and fully responsible for resolving the interference unless the operation of one station is recommended by a frequency coordinator and the operation of the other station is not. In that case, the licensee of the noncoordinated repeater has primary responsibility to resolve the interference.
(d) A repeater may be automatically controlled.
(e) Ancillary functions of a repeater that are available to users on the input channel are not considered remotely controlled functions of the station. Limiting the use of a repeater to only certain user stations is permissible.
(f) [Reserved]
(g) The control operator of a repeater that retransmits inadvertently communications that violate the rules in this Part is not accountable for the violative communications.
(h) The provisions of this paragraph do not apply to repeaters that transmit on the 1.2 cm or shorter wavelength bands. Before establishing a repeater within 16 km (10 miles) of the Arecibo Observatory or before changing the transmitting frequency, transmitter power, antenna height or directivity of an existing repeater, the station licensee must give written notification thereof

to the Interference Office, Arecibo Observatory, HC3 Box 53995, Arecibo, Puerto Rico 00612, in writing or electronically, of the technical parameters of the proposal. Licensees who choose to transmit information electronically should e-mail to: **prcz@naic. edu**.

1. The notification shall state the geographical coordinates of the antenna (NAD-83 datum), antenna height above mean sea level (AMSL), antenna center of radiation above ground level (AGL), antenna directivity and gain, proposed frequency and FCC Rule Part, type of emission, effective radiated power, and whether the proposed use is itinerant. Licensees may wish to consult interference guidelines provided by Cornell University.

2. If an objection to the proposed operation is received by the FCC from the Arecibo Observatory, Arecibo, Puerto Rico, within 20 days from the date of notification, the FCC will consider all aspects of the problem and take whatever action is deemed appropriate. The licensee will be required to make reasonable efforts in order to resolve or mitigate any potential interference problem with the Arecibo Observatory.

§97.207 Space station.
(a) Any amateur station may be a space station. A holder of any class operator license may be the control operator of a space station, subject to the privileges of the class of operator license held by the control operator.
(b) A space station must be capable of effecting a cessation of transmissions by telecommand whenever such cessation is ordered by the FCC.
(c) The following frequency bands and segments are authorized to space stations:
    (1) The 17 m, 15 m, 12 m and 10 m bands, 6 mm, 4 mm, 2 mm and 1 mm bands; and
    (2) The 7.0-7.1 MHz, 14.00-14.25 MHz, 144-146 MHz, 435-438 MHz, 1260-1270 MHz and 2400-2450 MHz, 3.40-3.41 GHz, 5.83-5.85 GHz, 10.45-10.50 GHz and 24.00-24.05 GHz segments.
(d) A space station may automatically retransmit the radio signals of Earth stations and other space stations.
(e) A space station may transmit one-way communications.
(f) Space telemetry transmissions may consist of specially coded messages intended to facilitate communications or related to the function of the spacecraft.
(g) The license grantee of each space station must make the following written notifications to the International Bureau, FCC, Washington, DC 20554.
    (1) A pre-space notification within 30 days after the date of launch vehicle determination, but no later than 90 days before integration of the space station into the launch vehicle. The notification must be in accordance with the provisions of Articles 9 and 11 of the International Telecommunication Union (ITU) Radio Regulations and must specify the information required by Appendix 4 and Resolution No. 642 of the ITU Radio Regulations. The notification must also include a description of the design and operational strategies that the space station will use to mitigate orbital debris, including the following information:
        (i) A statement that the space station licensee has assessed and limited the amount of debris released in a planned manner during normal operations, and has assessed and limited the probability of the space station becoming a source of debris by collisions with small debris or meteoroids that could cause loss of control and prevent post-mission disposal;
        (ii) A statement that the space station licensee has assessed and limited the probability of accidental explosions during and after completion of mission operations. This statement must include a demonstration that debris generation will not result from the conversion of energy sources on board the spacecraft into energy that fragments the spacecraft. Energy sources include chemical, pressure, and kinetic energy. This demonstration should address whether stored energy will be removed at the spacecraft's end of life, by depleting residual fuel and leaving all fuel line valves open, venting any pressurized system, leaving all batteries in a permanent discharge state, and removing any remaining source of stored energy, or through other equivalent procedures specifically disclosed in the application;
        (iii) A statement that the space station licensee has assessed and limited the probability of the space station becoming a source of debris by collisions with large debris or other operational space stations. Where a space station will be launched into a low-Earth orbit that is identical, or very similar, to an orbit used by other space stations, the statement must include an analysis of the potential risk of collision and a description of what measures the space station operator plans to take to avoid in-orbit collisions. If the space station licensee is relying on coordination with another system, the statement must indicate what steps have been taken to contact, and ascertain the likelihood of successful coordination of physical operations with, the other system. The statement must disclose the accuracy—if any—with which orbital parameters of non-geostationary satellite orbit space stations will be maintained, including apogee, perigee, inclination, and the right ascension of the ascending node(s). In the event that a system is not able to maintain orbital tolerances, i.e., it lacks a propulsion system for orbital maintenance, that fact should be included in the debris mitigation disclosure. Such systems must also indicate the anticipated evolution over time of the orbit of the proposed satellite or

satellites. Where a space station requests the assignment of a geostationary-Earth orbit location, it must assess whether there are any known satellites located at, or reasonably expected to be located at, the requested orbital location, or assigned in the vicinity of that location, such that the station keeping volumes of the respective satellites might overlap. If so, the statement must include a statement as to the identities of those parties and the measures that will be taken to prevent collisions;

(iv) A statement detailing the post-mission disposal plans for the space station at end of life, including the quantity of fuel—if any— that will be reserved for post-mission disposal maneuvers. For geostationary-Earth orbit space stations, the statement must disclose the altitude selected for a post-mission disposal orbit and the calculations that are used in deriving the disposal altitude. The statement must also include a casualty risk assessment if planned post- mission disposal involves atmospheric re-entry of the space station. In general, an assessment should include an estimate as to whether portions of the spacecraft will survive re-entry and reach the surface of the Earth, as well as an estimate of the resulting probability of human casualty.

(v) If any material item described in this notification changes before launch, a replacement pre-space notification shall be filed with the International Bureau no later than 90 days before integration of the space station into the launch vehicle.

(2) An in-space station notification is required no later than 7 days following initiation of space station transmissions. This notification must update the information contained in the pre-space notification.

(3) A post-space station notification is required no later than 3 months after termination of the space station transmissions. When termination of transmissions is ordered by the FCC, the notification is required no later than 24 hours after termination of transmissions.

§97.209 Earth station.

(a) Any amateur station may be an Earth station. A holder of any class operator license may be the control operator of an Earth station, subject to the privileges of the class of operator license held by the control operator.

(b) The following frequency bands and segments are authorized to Earth stations:

(1) The 17 m, 15 m, 12 m and 10 m bands, 6 mm, 4 mm, 2 mm and 1 mm bands; and

(2) The 7.0-7.1 MHz, 14.00-14.25 MHz, 144-146 MHz, 435-438 MHz, 1260-1270 MHz and 2400-2450 MHz, 3.40-3.41 GHz, 5.65-5.67 GHz, 10.45-10.50 GHz and 24.00-24.05 GHz segments.

§97.211 Space Telecommand station.

(a) Any amateur station designated by the licensee of a space station is eligible to transmit as a telecommand station for that space station, subject to the privileges of the class of operator license held by the control operator.

(b) A telecommand station may transmit special codes intended to obscure the meaning of telecommand messages to the station in space operation.

(c) The following frequency bands and segments are authorized to telecommand stations:

(1) The 17 m, 15 m, 12 m and 10 m bands, 6 mm, 4 mm, 2 mm and 1 mm bands; and

(2) The 7.0-7.1 MHz, 14.00-14.25 MHz, 144-146 MHz, 435-438 MHz, 1260-1270 MHz and 2400-2450 MHz, 3.40-3.41 GHz, 5.65-5.67 GHz, 10.45-10.50 GHz and 24.00-24.05 GHz segments.

(d) A telecommand station may transmit one-way communications.

§97.213 Telecommand of an amateur station.

An amateur station on or within 50 km of the Earth's surface may be under telecommand where:

(a) There is a radio or wireline control link between the control point and the station sufficient for the control operator to perform his/her duties. If radio, the control link must use an auxiliary station. A control link using a fiber optic cable or another telecommunication service is considered wireline.

(b) Provisions are incorporated to limit transmission by the station to a period of no more than 3 minutes in the event of malfunction in the control link.

(c) The station is protected against making, willfully or negligently, unauthorized transmissions.

(d) A photocopy of the station license and a label with the name, address, and telephone number of the station licensee and at least one designated control operator is posted in a conspicuous place at the station location.

§97.215 Telecommand of model craft.

An amateur station transmitting signals to control a model craft may be operated as follows:

(a) The station identification procedure is not required for transmissions directed only to the model craft, provided that a label indicating the station call sign and the station licensee's name and address is affixed to the station transmitter.

(b) The control signals are not considered codes or ciphers intended to obscure the meaning of the communication.

(c) The transmitter power must not exceed 1 W.

§97.217 Telemetry.
Telemetry transmitted by an amateur station on or within 50 km of the Earth's surface is not considered to be codes or ciphers intended to obscure the meaning of communications.

§97.219 Message forwarding system.
(a) Any amateur station may participate in a message forwarding system, subject to the privileges of the class of operator license held.
(b) For stations participating in a message forwarding system, the control operator of the station originating a message is primarily accountable for any violation of the rules in this Part contained in the message.
(c) Except as noted in paragraph (d) of this section, for stations participating in a message forwarding system, the control operators of forwarding stations that retransmit inadvertently communications that violate the rules in this Part are not accountable for the violative communications. They are, however, responsible for discontinuing such communications once they become aware of their presence.
(d) For stations participating in a message forwarding system, the control operator of the first forwarding station must:
    (1) Authenticate the identity of the station from which it accepts communication on behalf of the system; or
    (2) Accept accountability for any violation of the rules in this Part contained in messages it retransmits to the system.

§97.221 Automatically controlled digital station.
(a) This rule section does not apply to an auxiliary station, a beacon station, a repeater station, an earth station, a space station, or a space telecommand station.
(b) A station may be automatically controlled while transmitting a RTTY or data emission on the 6 m or shorter wavelength bands, and on the 28.120-28.189 MHz, 24.925-24.930 MHz, 21.090-21.100 MHz, 18.105- 18.110 MHz, 14.0950-14.0995 MHz, 14.1005-14.112 MHz, 10.140-10.150 MHz, 7.100-7.105 MHz, or 3.585-3.600 MHz segments.
(c) A station may be automatically controlled while transmitting a RTTY or data emission on any other frequency authorized for such emission types provided that:
    (1) The station is responding to interrogation by a station under local or remote control; and
    (2) No transmission from the automatically controlled station occupies a bandwidth of more than 500 Hz.

Subpart D—Technical Standards

§97.301 Authorized frequency bands.
The following transmitting frequency bands are available to an amateur station located within 50 km of the Earth's surface, within the specified ITU Region, and outside any area where the amateur service is regulated by any authority other than the FCC.
(a) For a station having a control operator who has been granted a Technician, Technician Plus, General, Advanced, or Amateur Extra Class operator license, who holds a CEPT radio amateur license, or who holds any class of IARP:

Wavelength band	ITU Region 1	ITU Region 2	ITU Region 3	Sharing requirements, see §97.303, paragraph:
VHF	MHz			
6 m	—	50-54	50-54	(a)
2 m	144-146	144-148	144-148	(a)
1.25 m	-	219-220	-	(a), (e)
-do-	—	222-225	—	(a)
UHF	MHz			
70 cm	430-440	420-450	420-450	(a), (b), (f)
33 cm	—	902-928	—	(a), (b), (g)
23 cm	1240-1300	1240-1300	1240-1300	(b), (h), (i)
13 cm	2300-2310	2300-2310	2300-2310	(a), (b), (j)
-do-	2390-2450	2390-2450	2390-2450	(a), (b), (j)
SHF	GHz			
9 cm	3.4-3.475	3.3-3.5	3.3-3.5	(a), (b), (k), (l)
5 cm	5.650-5.850	5.650-5.925	5.650-5.850	(a), (b), (m)
3 cm	10.00-10.50	10.00-10.50	10.00-10.50	(a), (c), (i), (n)
1.2 cm	24.00-24.25	24.00-24.25	24.00-24.25	(a), (b), (i), (o)
EHF	GHz			
6 mm	47.0-47.2	47.0-47.2	47.0-47.2	
4 mm	75.5-81.0	75.5-81.0	75.5-81.0	(b), (c), (h), (k), (r)
2.5 mm	122.25-123	122.25-123	122.25-123	(p)
2 mm	134-141	134-141	134-141	(b), (c), (h), (k)
1 mm	241-250above 275	241-250above 275	241-250above 275	(b), (c), (h), (k), (q).(k)

(b) For a station having a control operator who has been granted an Amateur Extra Class operator license, who holds a CEPT radio amateur license, or who holds a Class 1 IARP license:

Wavelength band	ITU Region 1	ITU Region 2	ITU Region 3	Sharing requirements, see §97.303, paragraph:
MF	kHz			
160 m	1810-1850	1800-2000	1800-2000	(a), (b), (c)
HF	MHz			
80 m	3.50-3.60	3.50-3.60	3.50-3.60	(a)
75 m	3.60-3.80	3.60-4.00	3.60-3.90	(a)
40 m	7.0-7.2	7.0-7.3	7.0-7.2	(a), (t)
30 m	10.10-10.15	10.10-10.15	10.10-10.15	(d)
20 m	14.00-14.35	14.00-14.35	14.00-14.35	
17 m	18.068-18.168	18.068-18.168	18.068-18.168	
15 m	21.00-21.45	21.00-21.45	21.00-21.45	
12 m	24.89-24.99	24.89-24.99	24.89-24.99	
10 m	28.0-29.7	28.0-29.7	28.0-29.7	

(c) For a station having a control operator who has been granted an operator license of Advanced Class:

Wavelength band	ITU Region 1	ITU Region 2	ITU Region 3	Sharing requirements, see §97.303, paragraph:
MF	kHz			
160 m	1810-1850	1800-2000	1800-2000	(a), (b), (c)
HF	MHz			
80 m	3.525-3.60	3.525-3.60	3.525-3.60	(a)
75 m	3.70-3.800	3.70-4.000	3.700-3.90	(a)
40 m	7.025-7.200	7.025-7.300	7.025-7.200	(a), (t)
30 m	10.10-10.15	10.10-10.15	10.10-10.15	(d)
20 m	14.025-14.150	14.025-14.150	14.025-14.150	
-do-	14.175-14.350	14.175-14.350	14.175-14.350	
17 m	18.068-18.168	18.068-18.168	18.068-18.168	
15 m	21.025-21.20	21.025-21.20	21.025-21.20	
-do-	21.225-21.45	21.225-21.45	21.225-21.45	

Wavelength band	ITU Region 1	ITU Region 2	ITU Region 3	
12 m	24.89-24.99	24.89-24.99	24.89-24.99	
10 m	28.0-29.7	28.0-29.7	28.0-29.7	

(d) For a station having a control operator who has been granted an operator license of General Class:

Wavelength band	ITU Region 1	ITU Region 2	ITU Region 3	Sharing requirements, see §97.303, paragraph:
MF	kHz			
160 m	1810-1850	1800-2000	1800-2000	(a), (b), (c)
HF	MHz			
80 m	3.525-3.60	3.525-3.60	3.525-3.60	(a)
75 m	—	3.80-4.00	3.80-3.90	(a)
40 m	7.025-7.125	7.025-7.125	7.025-7.125	(a)
-do-	—	7.175-7.300	—	(a)
30 m	10.10-10.15	10.10-10.15	10.10-10.15	(d)
20 m	14.025-14.150	14.025-14.150	14.025-14.150	
-do-	14.225-14.350	14.225-14.350	14.225-14.350	
17 m	18.068-18.168	18.068-18.168	18.068-18.168	
15 m	21.025-21.20	21.025-21.20	21.025-21.20	
-do-	21.275-21.45	21.275-21.45	21.275-21.45	
12 m	24.89-24.99	24.89-24.99	24.89-24.99	
10 m	28.0-29.7	28.0-29.7	28.0-29.7	

(e) For a station having a control operator who has been granted an operator license of Novice Class, Technician Class, or Technician Plus Class:

Wavelength band	ITU Region 1	ITU Region 2	ITU Region 3	Sharing requirements, see §97.303, paragraph:
HF	MHz			
80 m	3.525-3.60	3.525-3.60	3.525-3.60	(a)
40 m	7.025-7.075	7.025-7.125	7.025-7.075	(a)
15 m	21.025-21.20	21.025-21.20	21.025-21.20	
10 m	28.0-28.5	28.0-28.5	28.0-28.5	
VHF	MHz			
1.25 m	—	222-225	—	(a)
UHF	MHz			
23 cm	1270-1295	1270-1295	1270-1295	(h), (i)

§97.303 Frequency sharing requirements.

The following is a summary of the frequency sharing requirements that apply to amateur station transmissions on the frequency bands specified in §97.301 of this Part. (For each ITU Region, each frequency band allocated to the amateur service is designated as either a secondary service or a primary service. A station in a secondary service must not cause harmful interference to, and must accept interference from, stations in a primary service. See §§2.105 and 2.106 of the FCC Rules, *United States Table of Frequency Allocations* for complete requirements.)

(a) Where, in adjacent ITU Regions or sub-Regions, a band of frequencies is allocated to different services of the same category (i.e., primary or secondary allocations), the basic principle is the equality of right to operate. Accordingly, stations of each service in one Region or sub-Region must operate so as not to cause harmful interference to any service of the same or higher category in the other Regions or sub-Regions. (See ITU Radio Regulations, edition of 2004, No. 4.8.)

(b) No amateur station transmitting in the 1900-2000 kHz segment, the 70 cm band, the 33 cm band, the 23 cm band, the 13 cm band, the 9 cm band, the 5 cm band, the 3 cm band, the 24.05-24.25 GHz segment, the 76-77.5 GHz segment, the 78-81 GHz segment, the 136-141 GHz segment, and the 241-248 GHz segment shall not cause harmful interference to, nor is protected from interference due to the operation of, the Federal radiolocation service.

(c) No amateur station transmitting in the 1900-2000 kHz segment, the 3 cm band, the 76-77.5 GHz segment, the 78-81 GHz segment, the 136- 141 GHz segment, and the 241-248 GHz segment shall cause harmful interference to, nor is protected from interference due to the operation of, stations in the non-Federal radiolocation service.

(d) No amateur station transmitting in the 30 meter band shall cause harmful interference to stations authorized by other nations in the fixed service. The licensee of the amateur station must make all necessary adjustments, including termination of transmissions, if harmful interference is caused.

(e) In the 1.25 m band:

    (1) Use of the 219-220 MHz segment is limited to amateur stations participating, as forwarding stations, in point-to-point fixed digital message forwarding systems, including intercity packet backbone networks. It is not available for other purposes.

    (2) No amateur station transmitting in the 219-220 MHz segment shall cause harmful interference to, nor is protected from interference due to operation of Automated Maritime Telecommunications Systems (AMTS), television broadcasting on channels 11 and 13, 218-219 MHz Service systems, Land Mobile Services systems, or any other service having a primary allocation in or adjacent to the band.

(3) No amateur station may transmit in the 219-220 MHz segment unless the licensee has given written notification of the station's specific geographic location for such transmissions in order to be incorporated into a data base that has been made available to the public. The notification must be given at least 30 days prior to making such transmissions. The notification must be given to:
The American Radio Relay League
225 Main Street
Newington, CT 06111-1494

(4) No amateur station may transmit in the 219-220 MHz segment from a location that is within 640 km of an AMTS Coast Station that uses frequencies in the 217-218/219-220 MHz AMTS bands unless the amateur station licensee has given written notification of the station's specific geographic location for such transmissions to the AMTS licensee. The notification must be given at least 30 days prior to making such transmissions. The location of AMTS Coast Stations using the 217-218/219-220 MHz channels may be obtained from either:
The American Radio Relay League
225 Main Street
Newington, CT 06111-1494
or
Interactive Systems, Inc.
Suite 1103
1601 North Kent Street
Arlington, VA 22209
Fax: (703) 812-8275
Phone: (703) 812-8270

(5) No amateur station may transmit in the 219-220 MHz segment from a location that is within 80 km of an AMTS Coast Station that uses frequencies in the 217-218/219-220 MHz AMTS bands unless that amateur station licensee holds written approval from that AMTS licensee. The location of AMTS Coast Stations using the 217-218/219-220 MHz channels may be obtained as noted in paragraph (e)(4) of this section.

(f) In the 70 cm band:

(1) No amateur station shall transmit from north of Line A in the 420-430 MHz segment.

(2) The 420-430 MHz segment is allocated to the amateur service in the United States on a secondary basis, and is allocated in the fixed and mobile (except aeronautical mobile) services in the International Table of allocations on a primary basis. No amateur station transmitting in this band shall cause harmful interference to, nor is protected from interference due to the operation of, stations authorized by other nations in the fixed and mobile (except aeronautical mobile) services.

(3) The 430-440 MHz segment is allocated to the amateur service on a secondary basis in ITU Regions 2 and 3. No amateur station transmitting in this band in ITU Regions 2 and 3 shall cause harmful interference to, nor is protected from interference due to the operation of, stations authorized by other nations in the radiolocation service. In ITU Region 1, the 430-440 MHz segment is allocated to the amateur service on a co-primary basis with the radiolocation service. As between these two services in this band in ITU Region 1, the basic principle that applies is the equality of right to operate. Amateur stations authorized by the United States and radiolocation stations authorized by other nations in ITU Region 1 shall operate so as not to cause harmful interference to each other.

(4) No amateur station transmitting in the 449.75-450.00 MHz segment shall cause interference to, nor is protected from interference due to the operation of stations in, the space operation and space research services.

(g) In the 33 cm band:

(1) In the States of Colorado and Wyoming, bounded by the area of latitude 39° N. to 42° N. and longitude 103° W. to 108° W., an amateur station may transmit in the 902 MHz to 928 MHz band only on the frequency segments 902.0-902.4, 902.6-904.3, 904.7- 925.3, 925.7-927.3, and 927.7-928.0 MHz. This band is allocated on a secondary basis to the amateur service subject to not causing harmful interference to, and not receiving any interference protection from, the operation of industrial, scientific and medical devices, automatic vehicle monitoring systems, or Government stations authorized in this band.

(2) No amateur station shall transmit from those portions of the States of Texas and New Mexico bounded on the south by latitude 31° 41' N, on the north by latitude 34° 30' N, on the east by longitude 104° 11' W, and on the west by longitude 107° 30' W.

(h) No amateur station transmitting in the 23 cm band, the 3.3-3.4 GHz segment, the 3 cm band, the 24.05-24.25 GHz segment, the 76-77.5 GHz segment, the 78-81 GHz segment, the 136-141 GHz segment, and the 241-248 GHz segment shall cause harmful interference to, nor is protected from interference due to the operation of, stations authorized by other nations in the radiolocation service.

(i) In the 23 cm band, no amateur station shall cause harmful interference to, nor is protected from interference due to the operation of, stations in the radionavigation-satellite service, the aeronautical radionavigation service, the Earth exploration-satellite service (active), or the space research service (active).

(j) In the 13 cm band:
: (1) The amateur service is allocated on a secondary basis in all ITU Regions. In ITU Region 1, no amateur station shall cause harmful interference to, and shall be not protected from interference due to the operation of, stations authorized by other nations in the fixed and mobile services. In ITU Regions 2 and 3, no amateur station shall cause harmful interference to, and shall not be protected from interference due to the operation of, stations authorized by other nations in the fixed, mobile and radiolocation services.
: (2) In the United States:
:: (i) The 2300-2305 MHz segment is allocated to the amateur service on a secondary basis. (Currently the 2300-2305 MHz segment is not allocated to any service on a primary basis.);
:: (ii) The 2305-2310 MHz segment is allocated to the amateur service on a secondary basis to the fixed, mobile, and radiolocation services;
:: (iii) The 2390-2417 MHz segment is allocated to the amateur service on a primary basis.
::: (A) The 2390-2395 MHz segment is shared with Federal and non-Federal Government mobile services on a co-equal basis. See 47 CFR 2.106, footnote US276.
::: (B) Amateur stations operating in the 2400-2417 MHz segment must accept harmful interference that may be caused by the proper operation of industrial, scientific and medical equipment.
:: (iv) The 2417-2450 MHz segment is allocated to the amateur service on a co-secondary basis with the Federal Government radiolocation service. Amateur stations operating within the 2417-2450 MHz segment must accept harmful interference that may be caused by the proper operation of industrial, scientific, and medical devices operating within the band.

(k) No amateur station transmitting in the following segments shall cause harmful interference to stations in the radio astronomy service: 3.332-3.339 GHz, 3.3458-3.3525 GHz, 76-77.5 GHz, 78-81 GHz, 136-141 GHz, 241-248 GHz, 275-323 GHz, 327-371 GHz, 388-424 GHz, 426-442 GHz, 453-510 GHz, 623-711 GHz, 795-909 GHz, and 926-945 GHz. No amateur station transmitting in following segments shall cause harmful interference to stations in the Earth exploration-satellite service (passive) and space research service (passive): 275-277 GHz, 294-306 GHz, 316-334 GHz, 342-349 GHz, 363-365 GHz, 371-389 GHz, 416-434 GHz, 442-444 GHz, 496-506 GHz, 546-568 GHz, 624-629 GHz, 634-654 GHz, 659- 661 GHz, 684-692 GHz, 730-732 GHz, 851-853 GHz, and 951-956 GHz.

(l) In the 9 cm band:
: (1) In ITU Regions 2 and 3, the 9 cm band is allocated to the amateur service on a secondary basis. In ITU Region 1, the segment 3.4- 3.475 GHz is allocated to the amateur service on a secondary basis for use only in Germany, Israel, and the United Kingdom.
: (2) In the United States, the 9 cm band is allocated to the amateur and non-Federal radiolocation services on a secondary basis.
: (3) In the 3.4-3.5 GHz segment, no amateur station shall cause harmful interference to, nor is protected from interference due to the operation of, stations in the fixed and fixed-satellite services.
: (4) In the 3.4-3.5 GHz segment, no amateur station shall cause harmful interference to, nor is protected from interference due to the operation of, stations authorized by other nations in the fixed and fixed-satellite service.

(m) In the 5 cm band:
: (1) In the 5.650-5.725 GHz segment, the amateur service is allocated in all ITU Regions on a co-secondary basis with the space research (deep space) service.
: (2) In the 5.725-5.850 GHz segment, the amateur service is allocated in all ITU Regions on a secondary basis. No amateur station shall cause harmful interference to, nor is protected from interference due to the operation of, stations authorized by other nations in the fixed-satellite service in ITU Region 1.
: (3) No amateur station transmitting in the 5.725-5.875 GHz segment is protected from interference due to the operation of industrial, scientific and medical devices operating on 5.8 GHz.
: (4) In the 5.650-5.850 GHz segment, no amateur station shall cause harmful interference to, nor is protected from interference due to the operation of, stations authorized by other nations in the radiolocation service.
: (5) In the 5.850-5.925 GHz segment, the amateur service is allocated in ITU Region 2 on a co-secondary basis with the radiolocation service. In the United States, the segment is allocated to the amateur service on a secondary basis to the non-Government fixed-satellite service. No amateur station shall cause harmful interference to, nor is protected from interference due to the operation of, stations authorized by other nations in the fixed, fixed-satellite and mobile services. No amateur station shall cause harmful interference to, nor is protected from interference due to the operation of, stations in the non-Government fixed-satellite service.

(n) In the 3 cm band:
: (1) In the United States, the 3 cm band is allocated to the amateur service on a co-secondary basis with the non-government radiolocation service.
: (2) In the 10.00-10.45 GHz segment in ITU Regions 1 and 3, no amateur station shall cause interference to, nor is protected from interference due to the operation of, stations authorized by other nations in the fixed and mobile services.

(o) No amateur station transmitting in the 1.2 cm band is protected from interference due to the operation of industrial, scientific and medical devices on 24.125 GHz. In the United States, the 24.05-24.25 GHz segment is allocated to the amateur service on a co-secondary basis with the non-government radiolocation and Government and non-government Earth exploration-satellite (active) services.

(p) The 2.5 mm band is allocated to the amateur service on a secondary basis. No amateur station transmitting in this band shall cause harmful interference to, nor is protected from interference due to the operation of, stations in the fixed, inter-satellite and mobile services.

(q) No amateur station transmitting in the 244-246 GHz segment of the 1 mm band is protected from interference due to the operation of industrial, scientific and medical devices on 245 GHz.

(r) In the 4 mm band:

>(1) Authorization of the 76-77 GHz segment of the 4 mm band for amateur station transmissions is suspended until such time that the Commission may determine that amateur station transmissions in this segment will not pose a safety threat to vehicle radar systems operating in this segment.

>(2) In places where the amateur service is regulated by the FCC, the 77.5-78 GHz segment is allocated to the amateur service and the amateur -satellite service on a co-primary basis with radiolocation services.

>(3) No amateur or amateur-satellite station transmitting in the 75.5-76 GHz segment shall cause interference to, nor is protected from, interference due to the operation of stations in the fixed service. After January 1, 2006, the 75.5-76 GHz segment is no longer allocated to the amateur service or to the amateur-satellite service.

(s) An amateur station having an operator holding a General, Advanced or Amateur Extra Class license may only transmit single sideband, suppressed carrier, (emission type 2K8J3E) upper sideband on the channels 5332 kHz, 5348 kHz, 5368 kHz, 5373 kHz, and 5405 kHz. Amateur stations shall ensure that their transmission occupies only the 2.8 kHz centered around each of these frequencies. Transmissions shall not exceed an effective radiated power (e.r.p.) of 50 W PEP. For the purpose of computing e.r.p. the transmitter PEP will be multiplied with the antenna gain relative to a dipole or equivalent calculation in decibels. A half wave dipole antenna will be presumed to have a gain of 0 dBd. Licenses using other antennas must maintain in their records either the manufacturer data on the antenna gain or calculations of the antenna gain. No amateur station may cause harmful interference to stations authorized in the mobile and fixed services; nor is any amateur station protected from interference due to the operation of any such station.

(t) (1) The 7-7.1 MHz segment is allocated to the amateur and amateur-satellite services on a primary and exclusive basis throughout the world, except that the 7-7.05 MHz segment is:

>(i) Additionally allocated to the fixed service on a primary basis in the countries listed in 47 CFR 2.106, footnote 5.140; and

>(ii) Alternatively allocated to the fixed service on a primary and exclusive basis (i.e., the segment 7-7.05 MHz is not allocated to the amateur service) in the countries listed in 47 CFR 2.106, footnote 5.141.

(2) The 7.1-7.2 MHz segment is allocated to the amateur service on an exclusive basis in Region 2. Until March 29, 2009, the 7.1-7.2 MHz segment is allocated to the amateur and broadcasting services on a co- primary basis in Region 1 and Region 3 and the use of the 7.1-7.2 MHz segment by the amateur service shall not impose constraints on the broadcasting service intended for use within Region 1 and Region 3. After March 29, 2009, the 7.1-7.2 MHz segment is allocated to the amateur service on a primary and exclusive basis throughout the world, except that the 7.1-7.2 MHz segment is additionally allocated to the fixed and mobile except aeronautical mobile (R) services on a primary basis in the countries listed in 47 CFR 2.106, footnote 5.141B.

(3) The 7.2-7.3 MHz segment is allocated to the amateur service on an exclusive basis in Region 2 and to the broadcasting service on an exclusive basis in Region 1 and Region 3. The use of the 7.2-7.3 MHz segment in Region 2 by the amateur service shall not impose constraints on the broadcasting service intended for use within Region 1 and Region 3.

§97.305 Authorized emission types.

(a) Except as specified elsewhere in this part, an amateur station may transmit a CW emission on any frequency authorized to the control operator.

(b) A station may transmit a test emission on any frequency authorized to the control operator for brief periods for experimental purposes, except that no pulse modulation emission may be transmitted on any frequency where pulse is not specifically authorized and no SS modulation emission may be transmitted on any frequency where SS is not specifically authorized.

(c) A station may transmit the following emission types on the frequencies indicated, as authorized to the control operator, subject to the standards specified in §97.307(f) of this part.

Wavelength band	Frequencies	Emission Types Authorized §97.307(f), paragraph:	Standards, see
MF:			
160 m	Entire band	RTTY, data	(3)
-do-	-do-	Phone, image	(1), (2)
HF:			
80 m	Entire band	RTTY, data	(3), (9)
75 m	Entire band	Phone, image	(1), (2)
40 m	7.000-7.100 MHz	RTTY, data	(3), (9)
40 m	7.075-7.100 MHz	Phone, image	(1), (2), (9), (11)
40 m	7.100-7.125 MHz	RTTY, data	(3), (9)
40 m	7.125-7.300 MHz	Phone, image	(1), (2)
30 m	Entire band	RTTY, data	(3)
20 m	14.00-14.15 MHz	RTTY, data	(3)
-do-	14.15-14.35 MHz	Phone, image	(1), (2)
17 m	18.068-18.110 MHz	RTTY, data	(3)
-do-	18.110-18.168 MHz	Phone, image	(1), (2)
15 m	21.0-21.2 MHz	RTTY, data	(3), (9)
-do-	21.20-21.45 MHz	Phone, image	(1), (2)
12 m	24.89-24.93 MHz	RTTY, data	(3)
-do-	24.93-24.99 MHz	Phone, image	(1), (2)
10 m	28.0-28.3 MHz	RTTY, data	(4)
-do-	28.3-28.5 MHz	Phone, image	(1), (2), (10)
-do-	28.5-29.0 MHz	Phone, image	(1), (2)
-do-	29.0-29.7 MHz	Phone, image	(2)
VHF:			
6 m	50.1-51.0 MHz	MCW, phone, image, RTTY, data	(2), (5).
-do-	51.0-54.0 MHz	MCW, phone, image, RTTY, data, test	(2), (5), (8).
2 m	144.1-148.0 MHz	MCW, phone, image, RTTY, data, test	(2), (5), (8)
1.25 m	219-220 MHz	Data	(13)
-do-	222-225 MHz	RTTY, data, test MCW, phone, SS, image	(2), (6), (8)
UHF:			
70 cm	Entire band	MCW, phone, image, RTTY, data, SS, test	(6), (8)
33 cm	Entire band	MCW, phone, image, RTTY, data, SS, test, pulse	(7), (8), (12)
23 cm	Entire band	MCW, phone, image, RTTY, data, SS, test	(7), (8), (12)
13 cm	Entire band	MCW, phone, image, RTTY, data, SS, test, pulse	(7), (8), (12)
SHF:			
9 cm	Entire band	MCW, phone, image, RTTY, data, SS, test, pulse	(7), (8), (12)
5 cm	Entire band	MCW, phone, image, RTTY, data, SS, test, pulse	(7), (8), (12)
3 cm	Entire band	MCW, phone, image, RTTY, data, SS, test	(7), (8), (12)
1.2 cm	Entire band	MCW, phone, image, RTTY, data, SS, test, pulse	(7), (8), (12)
EHF:			
6 mm	Entire band	MCW, phone, image, RTTY, data, SS, test, pulse	(7), (8), (12)
4 mm	Entire band	MCW, phone, image, RTTY, data, SS, test, pulse	(7), (8), (12)
2.5 mm	Entire band	MCW, phone, image, RTTY, data, SS, test, pulse	(7), (8), (12)
2 mm	Entire band	MCW, phone, image, RTTY, data, SS, test, pulse	(7), (8), (12)
1 mm	Entire band	MCW, phone, image, RTTY, data, SS, test, pulse	(7), (8), (12)
—	Above 300 GHz	MCW, phone, image, RTTY, data, SS, test, pulse	(7), (8), (12)

§97.307 Emission standards.

(a) No amateur station transmission shall occupy more bandwidth than necessary for the information rate and emission type being transmitted, in accordance with good amateur practice.

(b) Emissions resulting from modulation must be confined to the band or segment available to the control operator. Emissions outside the necessary bandwidth must not cause splatter or keyclick interference to operations on adjacent frequencies.

(c) All spurious emissions from a station transmitter must be reduced to the greatest extent practicable. If any spurious emission, including chassis or power line radiation, causes harmful interference to the reception of another radio station, the licensee of the interfering amateur station is required to take steps to eliminate the interference, in accordance with good engineering practice.

(d) For transmitters installed after January 1, 2003, the mean power of any spurious emission from a station transmitter or external RF amplifier transmitting on a frequency below 30 MHz must be at least 43 dB below the mean power of the fundamental emission. For transmissions installed on or before January 1, 2003, the mean power of any spurious emission from a station transmitter or external RF power amplifier transmitting on a frequency below 30 MHz must not exceed 50 mW and must be at least 40 dB below the mean power of the fundamental emission. For a transmitter of mean power less than 5W installed on or before January 1, 2003, the attenuation must be at least 30 dB. A transmitter built before April 15, 1977, or first marketed before January 1, 1978, is exempt from this requirement.

(e) The mean power of any spurious emission from a station transmitter or external RF power amplifier transmitting on a frequency between 30-225 MHz must be at least 60 dB below the mean power of the fundamental. For a transmitter having a mean power of 25 W or less, the mean power of any spurious emission supplied to the antenna transmission line must not exceed 25 µW and must be at least 40 dB below the mean power of the fundamental emission, but need not be reduced below the power of 10 µW. A transmitter built before April 15, 1977, or first marketed before January 1, 1978, is exempt from this requirement.

(f) The following standards and limitations apply to transmissions on the frequencies specified in §97.305(c) of this Part.

    (1) No angle-modulated emission may have a modulation index greater than 1 at the highest modulation frequency.

    (2) No non-phone emission shall exceed the bandwidth of a communications quality phone emission of the same modulation type. The total bandwidth of an independent sideband emission (having B as the first symbol), or a multiplexed image and phone emission, shall not exceed that of a communications quality A3E emission.

    (3) Only a RTTY or data emission using a specified digital code listed in §97.309(a) of this Part may be transmitted. The symbol rate must not exceed 300 bauds, or for frequency-shift keying, the frequency shift between mark and space must not exceed 1 kHz.

    (4) Only a RTTY or data emission using a specified digital code listed in §97.309(a) of this Part may be transmitted. The symbol rate must not exceed 1200 bauds. For frequency-shift keying, the frequency shift between mark and space must not exceed 1 kHz.

    (5) A RTTY, data or multiplexed emission using a specified digital code listed in §97.309(a) of this Part may be transmitted. The symbol rate must not exceed 19.6 kilobauds. A RTTY, data or multiplexed emission using an unspecified digital code under the limitations listed in §97.309(b) of this Part also may be transmitted. The authorized bandwidth is 20 kHz.

    (6) A RTTY, data or multiplexed emission using a specified digital code listed in §97.309(a) of this Part may be transmitted. The symbol rate must not exceed 56 kilobauds. A RTTY, data or multiplexed emission using an unspecified digital code under the limitations listed in §97.309(b) of this Part also may be transmitted. The authorized bandwidth is 100 kHz.

    (7) A RTTY, data or multiplexed emission using a specified digital code listed in §97.309(a) of this Part or an unspecified digital code under the limitations listed in §97.309(b) of this Part may be transmitted.

    (8) A RTTY or data emission having designators with A, B, C, D, E, F, G, H, J or R as the first symbol; 1, 2, 7 or 9 as the second symbol; and D or W as the third symbol is also authorized.

    (9) A station having a control operator holding a Novice or Technician Class operator license may only transmit a CW emission using the international Morse code.

    (10) A station having a control operator holding a Novice Class operator license or a Technician Class operator license may only transmit a CW emission using the international Morse code or phone emissions J3E and R3E.

    (11) Phone and image emissions may be transmitted only by stations located in ITU Regions 1 and 3, and by stations located within ITU Region 2 that are west of 130° West longitude or south of 20° North latitude.

    (12) Emission F8E may be transmitted.

    (13) A data emission using an unspecified digital code under the limitations listed in §97.309(b) of this Part also may be transmitted. The authorized bandwidth is 100 kHz.

§97.309 RTTY and data emission codes.

(a) Where authorized by §97.305(c) and 97.307(f) of this Part, an amateur station may transmit a RTTY or data emission using the following specified digital codes:

    (1) The 5-unit, start-stop, International Telegraph Alphabet No. 2, code defined in ITU-T Recommendation F.1, Division C (commonly known as "Baudot").

    (2) The 7-unit code specified in ITU-R Recommendations M.476-5 and M.625-3 (commonly known as "AMTOR").

    (3) The 7-unit, International Alphabet No. 5, code defined in ITU-T Recommendation T.50 (commonly known as "ASCII").

    (4) An amateur station transmitting a RTTY or data emission using a digital code specified in this paragraph may use any technique whose technical characteristics have been documented publicly, such as CLOVER, G-TOR, or PacTOR, for the purpose of facilitating communications.

(b) Where authorized by §§ 97.305(c) and 97.307(f) of this part, a station may transmit a RTTY or data emission using an unspecified digital code, except to a station in a country with which the United States does not have an agreement permitting the code to be used. RTTY and data emissions using unspecified digital codes must not be transmitted for the purpose of obscuring the meaning of any communication. When deemed necessary by a District Director to assure compliance with the FCC Rules, a station must:

(1) Cease the transmission using the unspecified digital code;
(2) Restrict transmissions of any digital code to the extent instructed;
(3) Maintain a record, convertible to the original information, of all digital communications transmitted.

§97.311 SS emission types.

(a) SS emission transmissions by an amateur station are authorized only for communications between points within areas where the amateur service is regulated by the FCC and between an area where the amateur service is regulated by the FCC and an amateur station in another country that permits such communications. SS emission transmissions must not be used for the purpose of obscuring the meaning of any communication.

(b) A station transmitting SS emissions must not cause harmful interference to stations employing other authorized emissions, and must accept all interference caused by stations employing other authorized emissions.

(c) When deemed necessary by a District Director to assure compliance with this Part, a station licensee must:
(1) Cease SS emission transmissions;
(2) Restrict SS emission transmissions to the extent instructed; and
(3) Maintain a record, convertible to the original information (voice, text, image, etc.) of all spread spectrum communications transmitted.

(d) The transmitter power must not exceed 100 W under any circumstances. If more than 1 W is used, automatic transmitter control shall limit output power to that which is required for the communication. This shall be determined by the use of the ratio, measured at the receiver, of the received energy per user data bit (E_b) to the sum of the received power spectral densities of noise (N_0) and co-channel interference (I_0). Average transmitter power over 1 W shall be automatically adjusted to maintain an $E_b/(N_0 + I_0)$ ratio of no more than 23 dB at the intended receiver.

§97.313 Transmitter power standards.

(a) An amateur station must use the minimum transmitter power necessary to carry out the desired communications.

(b) No station may transmit with a transmitter power exceeding 1.5 kW PEP.

(c) No station may transmit with a transmitter power exceeding 200 W PEP:
(1) On the 10.10-10.15 MHz segment; or
(2) On the 80, 40, 15 or 10-meter bands when the control operator is a Novice Class operator or Technician Class operator.

(d) No station may transmit with a transmitter power exceeding 25 W PEP on the VHF 1.25 m band when the control operator is a Novice operator.

(e) No station may transmit with a transmitter power exceeding 5 W PEP on the UHF 23 cm band when the control operator is a Novice operator.

(f) No station may transmit with a transmitter power exceeding 50 W PEP on the UHF 70 cm band from an area specified in footnote US7 to §2.106 of the FCC Rules, unless expressly authorized by the FCC after mutual agreement, on a case-by-case basis, between the District Director of the applicable field facility and the military area frequency coordinator at the applicable military base. An Earth station or telecommand station, however, may transmit on the 435-438 MHz segment with a maximum of 611 W effective radiated power (1 kW equivalent isotropically radiated power) without the authorization otherwise required. The transmitting antenna elevation angle between the lower half-power (-3 dB relative to the peak or antenna bore sight) point and the horizon must always be greater than 10°.

(g) No station may transmit with a transmitter power exceeding 50 W PEP on the 33 cm band from within 241 km of the boundaries of the White Sands Missile Range. Its boundaries are those portions of Texas and New Mexico bounded on the south by latitude 31° 41' North, on the east by longitude 104° 11' West, on the north by latitude 34° 30' North, and on the west by longitude 107° 30' West.

(h) No station may transmit with a transmitter power exceeding 50 W PEP on the 219-220 MHz segment of the 1.25 m band.

§97.315 Certification of external RF power amplifiers.

(a) Any external RF power amplifier (see § 2.815 of the FCC Rules) manufactured or imported for use at an amateur radio station must be certificated for use in the amateur service in accordance with subpart J of part 2 of the FCC Rules. No amplifier capable of operation below 144 MHz may be constructed or modified by a non-amateur service licensee without a grant of certification from the FCC.

(b) The requirement of paragraph (a) does not apply if one or more of the following conditions are met:
(1) The amplifier is constructed or modified by an amateur radio operator for use at an amateur station.
(2) The amplifier was manufactured before April 28, 1978, and has been issued a marketing waiver by the FCC, or the amplifier was purchased before April 28, 1978, by an amateur radio operator for use at that operator's station.
(3) The amplifier is sold to an amateur radio operator or to a dealer, the amplifier is purchased in used condition by a dealer, or the amplifier is sold to an amateur radio operator for use at that operator's station.

(c) Any external RF power amplifier appearing in the Commission's database as certificated for use in the amateur service may be marketed for use in the amateur service.

§97.317 Standards for certification of external RF power amplifiers.
(a) To receive a grant of certification, the amplifier must:
    (1) Satisfy the spurious emission standards of § 97.307 (d) or (e) of this part, as applicable, when the amplifier is operated at the lesser of 1.5 kW PEP or its full output power and when the amplifier is placed in the "standby" or "off" positions while connected to the transmitter.
    (2) Not be capable of amplifying the input RF power (driving signal) by more than 15 dB gain. Gain is defined as the ratio of the input RF power to the output RF power of the amplifier where both power measurements are expressed in peak envelope power or mean power.
    (3) Exhibit no amplification (0 dB gain) between 26 MHz and 28 MHz.
(b) Certification shall be denied when:
    (1) The Commission determines the amplifier can be used in services other than the Amateur Radio Service, or
    (2) The amplifier can be easily modified to operate on frequencies between 26 MHz and 28 MHz.

Subpart E—Providing Emergency Communications

§97.401 Operation during a disaster.
A station in, or within 92.6 km (50 nautical miles) of, Alaska may transmit emissions J3E and R3E on the channel at 5.1675 MHz (assigned frequency 5.1689 MHz) for emergency communications. The channel must be shared with stations licensed in the Alaska-Private Fixed Service. The transmitter power must not exceed 150 W PEP. A station in, or within 92.6 km of, Alaska may transmit communications for tests and training drills necessary to ensure the establishment, operation, and maintenance of emergency communication systems.

§97.403 Safety of life and protection of property.
No provision of these rules prevents the use by an amateur station of any means of radiocommunication at its disposal to provide essential communication needs in connection with the immediate safety of human life and immediate protection of property when normal communication systems are not available.

§97.405 Station in distress.
(a) No provision of these rules prevents the use by an amateur station in distress of any means at its disposal to attract attention, make known its condition and location, and obtain assistance.
(b) No provision of these rules prevents the use by a station, in the exceptional circumstances described in paragraph (a), of any means of radiocommunications at its disposal to assist a station in distress.

§97.407 Radio amateur civil emergency service.
(a) No station may transmit in RACES unless it is an FCC-licensed primary, club, or military recreation station and it is certified by a civil defense organization as registered with that organization, or it is an FCC-licensed RACES station. No person may be the control operator of a RACES station, or may be the control operator of an amateur station transmitting in RACES unless that person holds a FCC-issued amateur operator license and is certified by a civil defense organization as enrolled in that organization.
(b) The frequency bands and segments and emissions authorized to the control operator are available to stations transmitting communications in RACES on a shared basis with the amateur service. In the event of an emergency which necessitates invoking the President's War Emergency Powers under the provisions of section 706 of the Communications Act of 1934, as amended, 47 U.S.C. 606, RACES stations and amateur stations participating in RACES may only transmit on the frequency segments authorized pursuant to part 214 of this chapter.
(c) A RACES station may only communicate with:
    (1) Another RACES station;
    (2) An amateur station registered with a civil defense organization;
    (3) A United States Government station authorized by the responsible agency to communicate with RACES stations;
    (4) A station in a service regulated by the FCC whenever such communication is authorized by the FCC.
(d) An amateur station registered with a civil defense organization may only communicate with:
    (1) A RACES station licensed to the civil defense organization with which the amateur station is registered;
    (2) The following stations upon authorization of the responsible civil defense official for the organization with which the amateur station is registered:
        (i) A RACES station licensed to another civil defense organization;
        (ii) An amateur station registered with the same or another civil defense organization;
        (iii) A United States Government station authorized by the responsible agency to communicate with RACES stations; and
        (iv) A station in a service regulated by the FCC whenever such communication is authorized by the FCC.
(e) All communications transmitted in RACES must be specifically authorized by the civil defense organization for the area served. Only civil defense communications of the following types may be transmitted:
    (1) Messages concerning impending or actual conditions jeopardizing the public safety, or affecting the national defense or security during periods of local, regional, or national civil emergencies;
    (2) Messages directly concerning the immediate safety of life of individuals, the immediate protection of property, maintenance of law and order, alleviation of human suffering and need, and the combating of armed attack or sabotage;
    (3) Messages directly concerning the accumulation and dissemination of public information or instructions to the civilian population essential to the activities of the civil defense organization or other authorized governmental or relief agencies; and
    (4) Communications for RACES training drills and tests necessary to ensure the establishment and maintenance of orderly and efficient operation of the RACES as ordered by the responsible civil defense organizations served. Such drills and tests may not exceed a total time of 1 hour per week. With the approval of the chief officer for emergency planning the applicable State, Commonwealth, District or territory, however, such tests and drills may be conducted for a period not to exceed 72 hours no more than twice in any calendar year.

Subpart F—Qualifying Examinations Systems

§97.501 Qualifying for an amateur operator license.
Each applicant must pass an examination for a new amateur operator license grant and for each change in operator class. Each applicant for the class of operator license grant specified below must pass, or otherwise receive examination credit for, the following examination elements:
(a) Amateur Extra Class operator: Elements 2, 3, and 4;
(b) General Class operator: Elements 2 and 3;
(c) Technician Class operator: Element 2.

§97.503 Element standards.
A written examination must be such as to prove that the examinee possesses the operational and technical qualifications required to perform properly the duties of an amateur service licensee. Each written examination must be comprised of a question set as follows:
> (a) Element 2: 35 questions concerning the privileges of a Technician Class operator license. The minimum passing score is 26 questions answered correctly.
> (b) Element 3: 35 questions concerning the privileges of a General Class operator license. The minimum passing score is 26 questions answered correctly.
> (c) Element 4: 50 questions concerning the privileges of an Amateur Extra Class operator license. The minimum passing score is 37 questions answered correctly.

§97.505 Element credit.
(a) The administering VEs must give credit as specified below to an examinee holding any of the following license grants or license documents:
> (1) An unexpired (or expired but within the grace period for renewal) FCC-granted Advanced Class operator license grant: Elements 2 and 3.
> (2) An unexpired (or expired but within the grace period for renewal) FCC-granted General Class operator license grant: Elements 2 and 3.
> (3) An unexpired (or expired but within the grace period for renewal) FCC-granted Technician or Technician Plus Class operator (including a Technician Class operator license granted before February 14, 1991) license grant: Element 2.
> (4) An expired FCC-issued Technician Class operator license document granted before March 21, 1987; Element 3.
> (5) A CSCE: Each element the CSCE indicates the examinee passed within the previous 365 days.

(b) No examination credit, except as herein provided, shall be allowed on the basis of holding or having held any other license grant or document.

§97.507 Preparing an examination.
(a) Each written question set administered to an examinee must be prepared by a VE holding an Amateur Extra Class operator license. A written question set may also be prepared for the following elements by a VE holding an operator license of the class indicated:
> (1) Element 3: Advanced Class operator.
> (2) Element 2: Advanced or General.

(b) Each question set administered to an examinee must utilize questions taken from the applicable question pool.
(c) Each written question set administered to an examinee for an amateur operator license must be prepared, or obtained from a supplier, by the administering VEs according to instructions from the coordinating VEC.

§97.509 Administering VE requirements.
(a) Each examination for an amateur operator license must be administered by a team of at least 3 VEs at an examination session coordinated by a VEC. The number of examinees at the session may be limited.
(b) Each administering VE must:
> (1) Be accredited by the coordinating VEC;
> (2) Be at least 18 years of age;
> (3) Be a person who holds an amateur operator license of the class specified below:
>> (i) Amateur Extra, Advanced or General Class in order to administer a Technician Class operator license examination;
>> (ii) Amateur Extra or Advanced Class in order to administer a General Class operator license examination;
>> (iii) Amateur Extra Class in order to administer an Amateur Extra Class operator license examination.
>
> (4) Not be a person whose grant of an amateur station license or amateur operator license has ever been revoked or suspended.

(c) Each administering VE must be present and observing the examinee throughout the entire examination. The administering VEs are responsible for the proper conduct and necessary supervision of each examination. The administering VEs must immediately terminate the examination upon failure of the examinee to comply with their instructions.
(d) No VE may administer an examination to his or her spouse, children, grandchildren, stepchildren, parents, grandparents, stepparents, brothers, sisters, stepbrothers, stepsisters, aunts, uncles, nieces, nephews, and in-laws.
(e) No VE may administer or certify any examination by fraudulent means or for monetary or other consideration including reimbursement in any amount in excess of that permitted. Violation of this provision may result in the revocation of the grant of the VE's amateur station license and the suspension of the grant of the VE's amateur operator license.
(f) No examination that has been compromised shall be administered to any examinee. The same question set may not be re-administered to the same examinee.
(g) Upon completion of each examination element, the administering VEs must immediately grade the examinee's answers. The administering VEs are responsible for determining the correctness of the examinee's answers.
(h) When the examinee is credited for all examination elements required for the operator license sought, 3 VEs must certify that the examinee is qualified for the license grant and that the VEs have complied with these administering VE requirements. The certifying VEs are jointly and individually accountable for the proper administration of each examination element reported. The certifying VEs may delegate to other qualified VEs their authority, but not their accountability, to administer individual elements of an examination.
(i) When the examinee does not score a passing grade on an examination element, the administering VEs must return the application document to the examinee and inform the examinee of the grade.
(j) The administering VEs must accommodate an examinee whose physical disabilities require a special examination procedure. The administering VEs may require a physician's certification indicating the nature of the disability before determining which, if any, special procedures must be used.
(k) The administering VEs must issue a CSCE to an examinee who scores a passing grade on an examination element.
(l) After the administration of a successful examination for an amateur operator license, the administering VEs must submit the application document to the coordinating VEC according to the coordinating VEC's instructions.

§97.511 Examinee conduct.
Each examinee must comply with the instructions given by the administering VEs.

§97.513 VE session manager requirements.
(a) A VE session manager may be selected by the VE team for each examination session. The VE session manager must be accredited as a VE by the same VEC that coordinates the examination session. The VE session manager may serve concurrently as an administering VE.
(b) The VE session manager may carry on liaison between the VE team and the coordinating VEC.
(c) The VE session manager may organize activities at an examination session.

§97.515 [Reserved]

§97.517 [Reserved]

§97.519 Coordinating examination sessions.
(a) A VEC must coordinate the efforts of VEs in preparing and administering examinations.
(b) At the completion of each examination session, the coordinating VEC must collect applicant information and test results from the administering VEs. The coordinating VEC must:
    (1) Screen collected information;
    (2) Resolve all discrepancies and verify that the VE's certifications are properly completed; and
    (3) For qualified examinees, forward electronically all required data to the FCC. All data forwarded must be retained for at least 15 months and must be made available to the FCC upon request.
(c) Each VEC must make any examination records available to the FCC, upon request.
(d) The FCC may:
    (1) Administer any examination element itself;
    (2) Readminister any examination element previously administered by VEs, either itself or under the supervision of a VEC or VEs designated by the FCC; or
    (3) Cancel the operator/primary station license of any licensee who fails to appear for readministration of an examination when directed by the FCC, or who does not successfully complete any required element that is readministered. In an instance of such cancellation, the person will be granted an operator/primary station license consistent with completed examination elements that have not been invalidated by not appearing for, or by failing, the examination upon readministration.

§97.521 VEC qualifications.
No organization may serve as a VEC unless it has entered into a written agreement with the FCC. The VEC must abide by the terms of the agreement. In order to be eligible to be a VEC, the entity must:
(a) Be an organization that exists for the purpose of furthering the amateur service;
(b) Be capable of serving as a VEC in at least the VEC region (see Appendix 2) proposed;
(c) Agree to coordinate examinations for any class of amateur operator license;
(d) Agree to assure that, for any examination, every examinee qualified under these rules is registered without regard to race, sex, religion, national origin or membership (or lack thereof) in any amateur service organization.

§97.523 Question pools.
All VECs must cooperate in maintaining one question pool for each written examination element. Each question pool must contain at least 10 times the number of questions required for a single examination. Each question pool must be published and made available to the public prior to its use for making a question set. Each question on each VEC question pool must be prepared by a VE holding the required FCC-issued operator license. See §97.507(a) of this Part.

§97.525 Accrediting VEs.
(a) No VEC may accredit a person as a VE if:
    (1) The person does not meet minimum VE statutory qualifications or minimum qualifications as prescribed by this Part;
    (2) The FCC does not accept the voluntary and uncompensated services of the person;
    (3) The VEC determines that the person is not competent to perform the VE functions; or
    (4) The VEC determines that questions of the person's integrity or honesty could compromise the examinations.
(b) Each VEC must seek a broad representation of amateur operators to be VEs. No VEC may discriminate in accrediting VEs on the basis of race, sex, religion or national origin; nor on the basis of membership (or lack thereof) in an amateur service organization; nor on the basis of the person accepting or declining to accept reimbursement.

§97.527 Reimbursement for expenses.
VEs and VECs may be reimbursed by examinees for out-of-pocket expenses incurred in preparing, processing, administering, or coordinating an examination for an amateur operator license.

Appendix 1

Places Where the Amateur Service is Regulated by the FCC

In ITU Region 2, the amateur service is regulated by the FCC within the territorial limits of the 50 United States, District of Columbia, Caribbean Insular areas [Commonwealth of Puerto Rico, United States Virgin Islands (50 islets and cays) and Navassa Island], and Johnston Island (Islets East, Johnston, North and Sand) and Midway Island (Islets Eastern and Sand) in the Pacific Insular areas.
In ITU Region 3, the amateur service is regulated by the FCC within the Pacific Insular territorial limits of American Samoa (seven islands), Baker Island, Commonwealth of Northern Mariannas Islands, Guam Island, Howland Island, Jarvis Island, Kingman Reef, Kure Island, Palmyra Island (more than 50 islets) and Wake Island (Islets Peale, Wake and Wilkes).

Appendix 2

VEC Regions
1. Connecticut, Maine, Massachusetts, New Hampshire, Rhode Island and Vermont.
2. New Jersey and New York.
3. Delaware, District of Columbia, Maryland and Pennsylvania.
4. Alabama, Florida, Georgia, Kentucky, North Carolina, South Carolina, Tennessee and Virginia.
5. Arkansas, Louisiana, Mississippi, New Mexico, Oklahoma and Texas.
6. California.
7. Arizona, Idaho, Montana, Nevada, Oregon, Utah, Washington and Wyoming.
8. Michigan, Ohio and West Virginia.
9. Illinois, Indiana and Wisconsin.
10. Colorado, Iowa, Kansas, Minnesota, Missouri, Nebraska, North Dakota and South Dakota.
11. Alaska.
12. Caribbean Insular areas.
13. Hawaii and Pacific Insular areas.

About the ARRL

The seed for Amateur Radio was planted in the 1890s, when Guglielmo Marconi began his experiments in wireless telegraphy. Soon he was joined by dozens, then hundreds, of others who were enthusiastic about sending and receiving messages through the air—some with a commercial interest, but others solely out of a love for this new communications medium. The United States government began licensing Amateur Radio operators in 1912.

By 1914, there were thousands of Amateur Radio operators—hams—in the United States. Hiram Percy Maxim, a leading Hartford, Connecticut inventor and industrialist, saw the need for an organization to band together this fledgling group of radio experimenters. In May 1914 he founded the American Radio Relay League (ARRL) to meet that need.

Today ARRL, with approximately 150,000 members, is the largest organization of radio amateurs in the United States. The ARRL is a not-for-profit organization that:
- promotes interest in Amateur Radio communications and experimentation
- represents US radio amateurs in legislative matters, and
- maintains fraternalism and a high standard of conduct among Amateur Radio operators.

At ARRL headquarters in the Hartford suburb of Newington, the staff helps serve the needs of members. ARRL is also International Secretariat for the International Amateur Radio Union, which is made up of similar societies in 150 countries around the world.

ARRL publishes the monthly journal *QST*, as well as newsletters and many publications covering all aspects of Amateur Radio. Its headquarters station, W1AW, transmits bulletins of interest to radio amateurs and Morse code practice sessions. The ARRL also coordinates an extensive field organization, which includes volunteers who provide technical information and other support services for radio amateurs as well as communications for public-service activities. In addition, ARRL represents US amateurs with the Federal Communications Commission and other government agencies in the US and abroad.

Membership in ARRL means much more than receiving *QST* each month. In addition to the services already described, ARRL offers membership services on a personal level, such as the ARRL Volunteer Examiner Coordinator Program and a QSL bureau.

Full ARRL membership (available only to licensed radio amateurs) gives you a voice in how the affairs of the organization are governed. ARRL policy is set by a Board of Directors (one from each of 15 Divisions). Each year, one-third of the ARRL Board of Directors stands for election by the full members they represent. The day-to-day operation of ARRL HQ is managed by an Executive Vice President and his staff.

No matter what aspect of Amateur Radio attracts you, ARRL membership is relevant and important. There would be no Amateur Radio as we know it today were it not for the ARRL. We would be happy to welcome you as a member! (An Amateur Radio license is not required for Associate Membership.) For more information about ARRL and answers to any questions you may have about Amateur Radio, write or call:

ARRL—The national association for Amateur Radio
225 Main Street
Newington CT 06111-1494
 Voice: 860-594-0200
    Fax: 860-594-0259
    E-mail: **hq@arrl.org**
    Internet: **www.arrl.org/**

Prospective new amateurs call (toll-free):
800-32-NEW HAM (800-326-3942)
You can also contact us via e-mail at **newham@arrl.org**
or check out *ARRLWeb* at **www.arrl.org/**

Notes

Notes

This book donated by the

ALGONQUIN AMATEUR RADIO CLUB

The club meets the second Thursday of the month except July and August in the Library of the Lt. Charles W. Whitcomb Middle school, Marlborough MA. at 7:30pm. Enter at the rear door #1. Library is on the left.

All are welcome.

for more information go to

www.n1em.org